DICTIONARY OF AVIATION

R.J. Hall
and
R.D. Campbell

St J
St James Press

Chicago and London

Published in North America by
St James Press
233 East Ontario Street
Chicago, Illinois 60611
U.S.A.

ISBN: 1–55862–106–7

Library of Congress Catalog Card Number: 90–63665

Printed in Great Britain by
BSP Professional Books 1991

CONTENTS

INTRODUCTION

The aviation industry, like many, has a wide range of technical terms, some of which vary from country to country. Even more confusing is the vast number of acronyms and abbreviations in use.

This book provides a compendium of words, terms, acronyms and abbreviations used in aviation. Many people engaged in, or connected with, the aviation industry will find it a useful reference work, particularly when seeking the definitions of aeronautical terms or abbreviations.

Some of the words defined in this book may be found in a standard dictionary, but are included here for one reason or another, particularly if their use in an aviation context differs from their usual one.

The book is divided into two sections: the first covers the definitions of aeronautical words, terms and phrases, while the second lists, in alphabetical order, a large number of acronyms and abbreviations. Where it is necessary for the reader to obtain more information than the brief explanation given in Abbreviations and Acronyms, reference should then be made to the definition in the Dictionary.

Where it is considered helpful to the reader, cross references are made between terms, and a number of useful definitions used by international aviation organisations, e.g. the International Civil Aviation Organisation (ICAO), the European Civil Aviation Conference (ECAC), etc., are included. These have been indicated by '[ICAO]' etc., in square brackets after the term.

In addition, and because the USA has such a large technical, industrial and operational influence on the aviation industry, a number of American aviation words and terms have been included. A second reason is that, due to the large number of American-manufactured aircraft used in the public transport and general aviation sectors, knowledge of American terminology is often a prerequisite for the safe operation of such aircraft. Certain FAA terms and ATC procedures have been given in order to acquaint the reader with the differences which exist between European and American airspace procedures. Some definitions have been expanded beyond what one might normally expect. This has been done for those items of information which the reader would normally have difficulty in finding from other sources, for example, the objectives and operating structure of the main international organisations and associations, particularly as their activities have a significant affect upon developments in aviation. Greater detail has also been given to new developments, such as cabin water spray systems to combat fuselage fires and thermal neutron analysis equipment used to detect explosives.

Aviation is a highly progressive industry, with new technology and terminology being continuously developed, but it is hoped that this publication will, for a while at least, keep the reader up-to-date with much of today's national and international aeronautical language.

Common Abbreviations

ATC	(Air Traffic Control) Air Transport Committee
EUROCONTROL	European Organisation for the Safety of Air Navigation
ICAO	International Civil Aviation Organisation
UK	United Kingdom
US	United States
WAFS	World Area Forecast System

The Dictionary

A

A-battery
Electric cell to heat the cathode filament in a valve.

abbreviated IFR flight plan
An authorisation by ATC requiring pilots to submit only that information needed for the purpose of ATC. It includes only a small portion of the usual IFR flight plan information. In certain instances, this may be only aircraft identification, location and pilot request. Other information may be requested if needed by ATC for separation/control purposes. It is frequently used by aircraft that are airborne and desire an instrument approach, or by aircraft on the ground that desire to climb to VMC on top.

abeam
An aircraft is 'abeam' a fix, point or object when that fix, point or object is approximately 90 degrees to the right or left of the aircraft track. Abeam indicates a general position rather than a precise point.

ab initio **training aircraft**
An aircraft designed and used for all basic pilot training.

ablation
In aeronautical usage — the erosion of the outer surface of a body travelling through the atmosphere at hypersonic speed.

abort
To terminate a pre-planned manoeuvre, e.g. an aborted takeoff.

abort drill
A sequence of actions forming the procedure to be adopted when aborting a manoeuvre.

above ground level
The height of an aircraft above ground level; also known as absolute altitude. May be determined by radio altimeter, or by subtracting the elevation of the ground from the flight altitude above mean sea level.

absolute ceiling
The altitude at which the maximum rate of climb is reduced to zero. Measured at a specified weight at 1 g in an international standard atmosphere.

absolute humidity
The mass of water vapour present in a unit volume of air, usually given in grammes per cubic metre.

absolute instability
A state of a layer within the atmosphere in which the vertical distribution of temperature is such that a parcel of air, if given an upward or downward push, will move away from its initial level without further outside force being applied.

absolute temperature scale   See KELVIN TEMPERATURE SCALE.

absolute vorticity   See VORTICITY.

absolute zero
$0°K$ (kelvin) $= -273.15°C$ $(-459.67°F)$.

accelerated stall
A stall entered from accelerated flight, g stall.

accelerate stop distance
The distance, for an aircraft, under the prevailing conditions, to accelerate to V_1 with all engines operating and then, with one engine failed to come to a stop, with the remaining engine(s) at idle or when available using reverse thrust.

acceleration
Rate of change of increasing velocity. As velocity is a vector quantity, acceleration can also be imparted by changing the flight trajectory without changing speed.

acceleration error
Erratic behaviour of a magnetic compass, caused by acceleration (or deceleration), when an aircraft is headed in a general east or west direction.

acceleration of gravity
$= 9.80665\,\mathrm{m.s}^{-2} = 32.174\,\mathrm{ft.s}^{-2}$

acceptance rate
Capacity threshold value for tactical air traffic flow management purposes, expressed in numbers of aircraft per given time period (normally a maximum of 60 minutes).

accident [ICAO]
An occurrence associated with the operation of an aircraft which takes places between the time any person boards the aircraft with the intention of flight until such time as all such persons have disembarked, in which:

(1) a person is fatally or seriously injured as a result of:

— being in the aircraft, or
— being in direct contact with any part of the aircraft, including parts which have become detached from the aircraft, or
— being in direct exposure to jet blast,

except when the injuries are from natural causes, self-inflicted or inflicted by other persons, or when the injuries are to stowaways hiding outside the areas normally available to the passengers and crew; or
(2) the aircraft sustains damage or structural failure which:

— adversely affects the structural strength, performance or flight characteristics of the aircraft, and

— would normally require major repair or replacement of the affected
component,

except for engine failure or damage, when the damage is limited to the engine,
its cowlings or accessories; or for damage limited to propellers, wing tips,
antennas, tyres, brakes, fairings, small dents or puncture holes in the aircraft
skin; or

(3) the aircraft is missing or is completely inaccessible.

Note 1: For statistical uniformity only, an injury resulting in death within thirty
days of the date of the accident is classified as a fatal injury by ICAO.

Note 2: An aircraft is considered to be missing when the official search has been
terminated and the wreckage has not been located.

accredited medical conclusion [ICAO]
The conclusion reached by one or more medical experts acceptable to the Licensing
Authority for the purposes of the case concerned, in consultation with flight
operators or other experts as necessary.

accuracy landing
A landing on a designated spot, used in pilot training, demonstration flying and
precision competitions.

acetylene gas
= C_2H_2 with a flame temperature of 3300°C when burned with oxygen.

acknowledge
Let me know that you have received and understood my message.

acoustic feedback
Self-oscillation in radio equipment created by a portion of the acoustic output
interfering with input.

acoustic spectrum
The range of frequencies emitted by a source of sound.

acoustic tube
An acoustic electric transducer fitted in flight crew headsets.

acrobatic flight [ICAO]
Manoeuvres intentionally performed by an aircraft involving an abrupt change in
its attitude, or an abnormal variation in speed.

active front
A front which produces considerable cloudiness and precipitation.

active runway   See RUNWAY IN USE.

actuator
A device imparting mechanical motion, usually over a specific rotary or linear
range.

additional services [US]
Advisory information provided by ATC that includes but is not limited to the following:

(1) Traffic advisories.
(2) Vectors, when requested by the pilot, to assist aircraft receiving traffic advisories to avoid observed traffic.
(3) Altitude deviation information of 300 ft or more from an assigned altitude as observed on a verified (reading correctly) automatic altitude readout (Mode C).
(4) Advisories that traffic is no longer a factor.
(5) Weather information.
(6) Weather assistance.
(7) Bird activity information.
(8) Holding pattern surveillance.

additional services are provided to the extent possible, contingent only upon the controller's capability to fit it into the performance of higher priority duties, and on the basis of limitations of the radar, volume of traffic, frequency congestion and controller workload.

additive
A substance added in small amounts to fuel, or other substance to improve performance.

address selective SSR
A recent development of SSR which permits interrogation of individual aircraft and exchange of data. Compatible with US Discrete Address Beacon System. (DABS)

adiabat
A line or curve on a temperature–pressure diagram along which a thermodynamic change takes place without the gain or loss of heat.

adiabatic
Without heat entering or leaving the system.

adiabatic lapse rate
The rate at which air temperature decreases as height increases above the earth's surface to the tropopause. See also DRY ADIABATIC LAPSE RATE and ENVIRONMENTAL LAPSE RATE.

adiabatic process
A thermodynamic process in which no heat is given to, or withdrawn from, the body of air concerned. Adiabatic changes of atmospheric temperature are those that occur only as a result of compression or expansion accompanying an increase or a decrease of atmospheric pressure. Such changes are also described as dynamic heating and cooling.

advanced
Adjective used to indicate latest technology or project philosophy.

advanced automation system
An ATC automation system developed by the FAA to support the next generation ATC systems in the USA.

advanced training aircraft
An aircraft used in pilot training and which is more powerful and complex than an *ab initio* aircraft.

advancing blade
On a helicopter in horizontal motion, a rotor blade moving forward against the relative airflow.

advection
(1) The movement of matter or energy from one place to another caused by a horizontal stream of gas.
(2) The horizontal transport of air or atmospheric properties. In meterology, sometimes referred to as the horizontal component of convection.

advection fog
Fog resulting from the transport of warm, humid air over a cold surface.

advise intentions
Tell me what you plan to do.

advisory route
A route within a flight information region along which air traffic advisory service is available.

Advisory Service
Advice and information provided by a facility to assist pilots in the safe conduct of flight and aircraft movement. See also ADDITIONAL SERVICES, AIRPORT ADVISORY SERVICE, EN-ROUTE FLIGHT ADVISORY SERVICE, RADAR ADVISORY, SAFETY ADVISORIES, TRAFFIC ADVISORIES.

aerial (antenna)
The part of a radio system which transmits or receives waves from the atmosphere or space.

aerial refuelling
A procedure used by the military to transfer fuel from one aircraft to another during flight.

aerial work
Specialised commercial aviation operations, not including air transport operations within the scope of ICAO Annex 6, Part 1, performed by aircraft, e.g. flying training, agriculture, construction, photography and surveying, etc.

aerobatics
Specific, usually standardised, manoeuvres performed in aircraft for the development of additional pilot skills, or for competitive purposes.

aerodrome
Any area of land or water designed, equipped, set apart, or commonly used for affording facilities for the landing and departure of aircraft and including any area or space, whether on the ground, on the roof of a building, or elsewhere, which is designed, equipped, or set apart for affording facilities for the landing and

departure of aircraft capable of descending or climbing vertically, but not including any area the use of which for the landing and departure of aircraft has been abandoned and not resumed.

aerodrome approach
That part of an instrument approach procedure commencing at the designated height over the radio aid to be used and ending when the aircraft has broken cloud.

aerodrome beacon
Aeronautical beacon used to indicate the location of an aerodrome from the air.

aerodrome climatological summary
Concise summary of specified meterological elements at an aerodrome, based on statistical data.

aerodrome climatological table
Table providing statistical data on the observed occurrence of one or more meteorological elements at an aerodrome.

aerodrome control radio station
A station providing radio communication between an aerodrome control tower and aircraft or mobile aeronautical stations.

aerodrome control service
Air traffic control service for aerodrome traffic.

aerodrome elevation
The elevation of the highest point of the landing area.

aerodrome flight information service
At aerodromes used by international general aviation, where the provision of aerodrome control is not yet justified, an aerodrome flight information service should be provided by an authorised person who is able to provide the pilots with relevant information, e.g. wind direction, runway in use, pressure setting, etc.

aerodrome location indicator
Four-letter identification code allocated to aerodromes. Used to simplify entries in flight plans and for other purposes.

aerodrome meterological office
An office, located at an aerodrome, designated to provide a meteorological service for international air navigation.

aerodrome operating minima
The limits of usability of an aerodrome for either takeoff or landing, usually expressed in terms of visibility or runway visual range, decision altitude/height, or minimum descent altitude/height and cloud conditions.

aerodrome operations [ICAO]
Assumed to commence at the beginning of the final approach segment and continue to the missed approach point for the approach phase. Also include ground operations after landing and prior to takeoff and encompass the takeoff ground roll and initial climb to the airfield departure procedure.

aerodrome reference point
The geographical location of the aerodrome and the centre of its traffic zone where an ATZ is established.

aerodrome taxi circuit
The specified path of aircraft on the manoeuvring area.

aerodrome traffic
All traffic on the manoeuvring area of an aerodrome and all aircraft flying in the vicinity of an aerodrome.
Note: An aircraft is in the vicinity of an aerodrome when it is in, entering, or leaving an aerodrome traffic circuit.

aerodrome traffic circuit
The specified path to be flown by aircraft operating in the vicinity of an aerodrome.

aerodrome traffic zone
An airspace of defined dimensions established around an aerodrome for the protection of aerodrome traffic.

aerodynamic centre
A point on the wing chord about which the resultant pitching moment of the aerodynamic forces acting upon it is constant. Generally about 25% of the wing chord measured from the leading edge.

aerodynamic coefficients
Non-dimensional coefficients for aerodynamic forces and moments.

aerodynamic efficiency
In general use applies to the ratio of the amount of lift to drag produced from an aerofoil. The greater the lift, or the less the drag, the greater the aerodynamic efficiency.

aerodynamic force
Reaction felt by the surface of a body as it moves through the air. Varies with changes in pressure and friction.

aerodynamic twisting
Change in angle of incidence along an aerofoil surface (aircraft wing or propeller) to vary the distribution of lift.

aerodynamic twisting movement
Generally propeller terminology. A consequence of the centre of pressure of the blade acting ahead of the centre of twist of the blade section. Thus during its rotation there is a tendency for the blade to twist about the propeller hub axis, creating a force which tries to turn the blade to a higher angle.

aerodynamics
The study of air in motion, or of moving bodies in air.

aeroelasticity
The interaction between aerodynamic forces and elastic structures causing change of shape or angular variations.

aerofoil
A surface shaped to produce more lift than drag when driven through the air.

aeromechanics
Science of motion and equilibrium of air and other gases, comprising aerodynamics and aerostatics.

aeromedicine
Medical study and treatment of disorders, diseases and disturbances resulting from, or associated with, atmospheric flight.

aerometeorograph
An aircraft instrument fitted to record various meteorological data, e.g. temperature, pressure, humidity.

aeronautical beacon
A visual navaid displaying flashes of white and/or coloured light to indicate the location of an airport, a heliport, a landmark, or in the United States a certain point of a Federal Airway in mountainous terrain or an obstruction.

Aeronautical Broadcasting Service
A broadcasting service intended for the transmission of information relating to air navigation.

aeronautical chart [ICAO]
A map used in air navigation containing all or part of the following: topographic features, hazards and obstructions, navigation aids, navigation routes, designated airspace and airports. Commonly used aeronautical charts are:

(1) Sectional charts, scale 1:500000. Designed for visual navigation of slow or medium speed aircraft. Topographical information on these charts features the portrayal of relief and a judicious selection of visual check points for VFR flight. Aeronautical information includes visual and radio aids to navigation, airports, controlled airspace, restricted areas, obstructions and related data.

(2) VFR terminal area charts, scale 1:250000. Depict terminal control area airspace that provides for the control or segregation of all the aircraft within. The chart depicts topographical information that includes visual and radio aids to navigation, airports, controlled airspace, restricted areas, obstructions and related data.

(3) World aeronautical charts (WAC), scale 1:1000000. A standard series of aeronautical charts covering land areas of the world, at a size and scale convenient for navigation by moderate speed aircraft. Topographical information includes cities and towns, principal roads, railroads, distinctive landmarks, drainage and relief. Aeronautical information includes visual and radio aids to navigation, airports, airways, restricted areas, obstructions and other pertinent data.

(4) En-route low altitude charts. Provide aeronautical information for en-route instrument navigation (IFR) in the low altitude stratum. Information includes the portrayal of airways, limits or controlled airspace, position identification and frequencies of radio aids, selected airports, minimum en-route and minimum obstruction clearance altitudes, airway distances, reporting points,

restricted areas and related data. Area charts that are a part of this series furnish terminal data at a larger scale in congested areas.

(5) En-route high altitude charts. Provide aeronautical information for en-route instrument navigation (IFR) in the high altitude stratum. Information includes the portrayal of jet routes, identification and frequencies of radio aids, selected airports, distances, time zones, special use airspace and related information.

(6) Instrument approach procedures (IAP) charts. Portray the aeronautical data that is required to execute an instrument approach to an airport. These charts depict the procedures, including all related data, and the airport diagram. Each procedure is designated for use with a specific type of electronic navigation system including ILS, NDB, RNAV, TACAN, and VOR. These charts are identified by the type of navigation aid(s) that provide final approach guidance.

(7) Standard instrument departure (SID) charts. Designed to expedite clearance delivery and to facilitate transition between takeoff and en-route operations. Each SID procedure is presented as a separate chart and may serve a single airport or more than one airport in a given geographical location.

(8) Standard terminal arrival route (STAR) charts. Designed to expedite air traffic control arrival route procedures and facilitate transition between en-route and instrument approach operation. Each STAR procedure is presented as a separate chart and may serve a single airport or more than one airport in a geographical location.

(9) Airport taxi charts. Designed to expedite the efficient and safe flow of ground traffic at an airport. These charts are identified by the official airport name.

Aeronautical Descriptive Climatological Memorandum
Description of the main climate features of concern to aviation for an area or an air route.

aeronautical fixed circuit
A circuit forming part of the aeronautical fixed service (AFS).

aeronautical fixed service
A telecommunications service between specified fixed points provided primarily for the safety of air navigation and for the regular, efficient and economical operation of air services.

aeronautical fixed station
A station in the aeronautical fixed service

aeronautical fixed telecommunications network
An integrated world-wide system of aeronautical fixed circuits provided, as part of the aeronautical fixed service, for the exchange of messages between the aeronautical fixed stations within the network.

aeronautical ground light
Any light specially provided as an aid to air navigation.

Aeronautical Information Circular
A notice containing information that does not qualify for the origination of a NOTAM or for inclusion in the AIP, but which relates to flight safety, air navigation, technical, administrative or legislative matters.

Aeronautical Information Publication
A publication issued by, or with the authority of, a State and containing aeronautical information of a lasting character essential to air navigation.

aeronautical information regulation and control
A system (and associated NOTAM) aimed at advance notification, based on common effective dates, of circumstances that necessitate changes in operating practices.

aeronautical meteorological station
A station designated to make observations and meteorological reports for use in international air navigation.

aeronautical mobile service
A radiocommunications service between aircraft stations and aeronautical stations, or between aircraft stations. AMS (OR) relates to flight communications primarily outside national/international air routes; AMS (R) relates to communications about safety and regularity of flight primarily along national/international air routes.

aeronautical radio navigation service
A radio navigation service intended for the benefit and for the safe operation of aircraft.

Note: The following Radio Regulations are quoted for purposes of reference and/or clarity in understanding the above definition of the aeronautical radio navigation service:

RR11 Radio navigation: Radio-determination used for the purpose of navigation, including obstruction warnings.

RR10 Radio-determination: The determination of the position, velocity and/or other characteristics of an object, or the obtaining of information relating to these parameters, by means of the propagation properties of radio waves.

aeronautical satellite
Space satellite providing communication and navigational services to aircraft.

aeronautical station
A land station in the aeronautical mobile service carrying on a service with aircraft stations. In certain instances an aeronautical station may be placed on board a ship or an earth satellite.

aeronautical telecommunication agency
An agency responsible for operating a station or stations in the aeronautical telecommunication service.

aeronautical telecommunication log
A record of the activities of an aeronautical telecommunication station.

aeronautical telecommunication service
A telecommunication service provided for any aeronautical purpose.

aeronautical telecommunication station
A station in the aeronautical telecommunication service.

aeronautics
Science of aircraft design, construction and operation.

aeroplane
A power-driven heavier-than-air aircraft, deriving its lift in flight chiefly from aerodynamic reactions on surfaces which remain fixed under given conditions of flight.

aeroplane flight manual
A manual, associated with the certificate of airworthiness, setting out the limitations within which the aeroplane is to be considered airworthy, and instructions and information necessary to the flight crew members for the safe operation of the aeroplane.

aeroplane reference field length
The minimum field length required for takeoff at maximum certificated takeoff mass, sea level, standard atmospheric conditions, still air and zero runway slope, as shown in the appropriate aeroplane flight manual prescribed by the certificating authority or equivalent data from the aeroplane manufacturer. Field length means balanced field length for aeroplanes, if applicable, or takeoff distance in other cases.

aeroplane system
An aeroplane system includes all elements of equipment necessary for the control and performance of a particular major function in question and other basic related aeroplane equipment such as that required to supply power for the equipment operation.

aerostat
Balloon or airship deriving its lift from the buoyancy of surrounding air.

aerostatics
Science of gases in equilibrium and equilibrium of balloons/airships in changing atmospheric conditions.

aerothermodynamics
Study of the thermodynamics of gases, particularly at high relative velocities.

aerotitis
A condition of the ear. Consistent with Boyle's law, air in the cavity of the middle ear expands and contracts with changes in atmospheric pressure. During altitude changes, if pressure in the ear is not readily equalised with the outside air pressure, pain will be experienced in the affected ear.

aero-tow flight time glider
The total time occupied in tow by an aeroplane from the moment the glider first moves for the purpose of taking off until the moment it is released from the tow device.

affirm
Signifies 'yes'.

afterburning
(1) The addition of fuel to the exhaust gases of a jet engine in order to increase the thrust.
(2) The irregular burning of the resident propellant in a rocket motor after the primary combustion has finished.

afterglow
(1) The glow after sunset caused by dust particles in the upper atmosphere scattering the sun's rays.
(2) Persistence of luminosity on a cathode ray tube screen or other luminescent equipment after the excitation has ceased.

AFTN communication centre
An AFTN station whose primary function is the relay or re-transmission of AFTN traffic from or to a number of other AFTN stations connected to it.

AFTN entry/exit points
Centres through which AFTN traffic entering and leaving an ICAO air navigation region should flow.

AFTN group
Three or more radio stations in the aeronautical fixed telecommunications network exchanging communications on the same radio frequency.

AFTN station
A station forming part of the aeronautical fixed telecommunications network (AFTN) and operating as such under the authority or control of a State.

agonic line
A line drawn on maps joining points of zero magnetic deviation.

agreed reporting point
A point specified in the route description of a flight plan and agreed between the operator and the air traffic services unit to serve as a reporting point for the flight concerned.

agricultural aircraft
An aircraft specifically designed or used for the purpose of spraying or spreading liquids or solid materials for agricultural purposes.

agricultural aviation
The section of general aviation which is concerned with the treatment of agriculture.

aileron
A horizontal control surface hinged to the mainplane.

air-air surveillance interrogation
An interrogation used by airborne collision avoidance system (ACAS) equipment to determine the positions of nearby Mode S-equipped aircraft and to exchange collision avoidance information air to air.

airborne
The period of flight between lifting off the earth's surface and landing.

airborne radio relay
Transmission and reception from an airborne radio station used for the purpose of increasing range.

air cargo
Mail, freight or goods carried by an aircraft.

air carrier
A person who undertakes directly by lease, or other arrangement, to engage in air transportation.

aircom
An ACARS compatible air–ground data link operated by SITA.

air commerce [US]
Interstate, overseas, or foreign air commerce, or the transportation of mail by aircraft, or any operation or navigation of aircraft which directly affects, or which may endanger safety in, interstate, overseas or foreign air commerce.

aircraft
Any machine that can derive support in the atmosphere from the reactions of the air other than the reactions of the air against the earth's surface.

aircraft avionics
Any electronic device, including radio, automatic flight control and instrument systems.

aircraft call sign
A group of alphanumeric characters used to identify an aircraft in air–ground communications.

aircraft category
(1) Classification of aircraft according to specified basic characteristics, e.g. aeroplane, glider, rotorcraft, free balloon.
Note: Categories of aircraft are defined in ICAO Annex 1.
(2) Category of aircraft in relation to instrument approach procedures. Aircraft performance differences have a direct effect on the air-space and visibility needed to perform certain manoeuvres such as circling approach, turning missed approach, and final approach descent and manoeuvring to land, including base and procedure turns. The most significant factor is speed. The following categories of typical aircraft are established, based on 1.3 times the stalling speed in the landing configuration at maximum certificated landing mass:

Category A: less than 91 knots IAS
        B: 91 knots or more but less than 121 knots IAS
        C: 121 knots or more but less than 141 knots IAS
        D: 141 knots or more but less than 166 knots IAS
        E: 166 knots or more but less than 211 knots IAS

aircraft certificate
Certificate issued in relation to an aircraft, used to indicate compliance with the appropriate requirements concerning aircraft type, airworthiness state, registration, etc.

aircraft classification number
A number expressing the relative effect of an aircraft on a pavement for a specified standard subgrade category.
Note: The aircraft classification number is calculated with respect to the centre of gravity (CG) position which yields the critical loading on the critical landing gear. Normally the most aft CG position appropriate to the maximum gross apron (ramp) mass is used to calculate the ACN. In exceptional cases the most forward CG position may result in the nose gear loading being more critical.

aircraft dispatcher [US]
A member of an airline operations team who is responsible for flight planning details that affect the safe conduct of the planned aircraft operation.

aircraft engine
An engine that is used, or intended to be used, for propelling aircraft. It includes turbosuperchargers, appurtenances and accessories necessary for its functioning, but does not include propellers.

aircraft equipment
Articles, other than stores and spare parts of a removable nature, for use on board an aircraft during flight, including first-aid and survival equipment.

aircraft identification
A group of letters, figures or a combination thereof which is either identical to, or the coded equivalent of, the aircraft call sign to be used in air–ground communications, and which is used to identify the aircraft in ground–air traffic services communications.

aircraft installation delay
The time elapsed between transmission and reception in a radio altimeter installation when the aircraft is in the touchdown position.

aircraft log books
Log books pertaining to a specific aircraft and maintained in respect of the aircraft airframe, engine and variable pitch propeller(s).

aircraft observation
The evaluation of one or more meteorological elements made from an aircraft in flight.

aircraft operating agency
The person, organisation or enterprise engaged in, or offering to engage in, an aircraft operation.

Aircraft Owners and Pilots Association [UK]
Objectives: The national and international representative body for UK general aviation, encompassing flying training organisations, aircraft owners, private pilots and professional pilots, and instructors employed in general aviation and aerial

work activities. The constitution and byelaws of its umbrella association, the International Council of Aircraft Owner and Pilot Associations, requires it to be a national autonomous body representing the interests of all classes of activities within general aviation and aerial work.

aircraft prepared for service weight
A fully equipped operational aeroplane, empty, i.e. without crew, fuel or payload.

aircraft stand
A designated area on an apron intended to be used for parking an aircraft.

aircraft station
A mobile station in the aeronautical mobile service, other than a survival craft station, located on board an aircraft.

aircraft-type
All aircraft of the same basic design including all modifications thereto except those modifications which result in a change in handling or flight characteristics.

air defense emergency [US]
A military emergency condition declared by a designated authority. This condition exists when an attack upon the continental United States, Alaska, Canada or US installations in Greenland by hostile aircraft or missiles is considered probable, is imminent, or is taking place.

air defense identification zone [US]
The area of airspace over land or water extending upward from the surface within which the ready identification, the location, and the control of aircraft are required in the interests of national security.

(1) Domestic air defense identification zone. An ADIZ within the United States along an international boundary of the United States.
(2) Coastal air defense identification zone. An ADIZ over the coastal waters of the United States.
(3) Distant early warning identification zone (DEWIZ). An ADIZ over the coastal waters of the State of Alaska.

air density
The mass density of the air in terms of weight per unit volume.

air-filed flight plan
A flight plan provided to an air traffic service unit by an aircraft during its flight.

airflow
The relative flow of air past a surface.

airframe
The fuselage, booms, nacelles, cowlings, fairings, aerofoil surfaces (including rotors but excluding propellers) and landing gear of an aircraft, and their accessories and controls.

air-ground communication
Two-way communication between aircraft and stations or locations on the surface of the earth.

air–ground control radio station
An aeronautical telecommunications station having primary responsibility for handling communications pertaining to the operation and control of aircraft in a given area.

air–ground facility   See REMOTE COMMUNICATIONS.

airline
As provided in Article 96 of the ICAO Convention, any air transport enterprise offering or operating a scheduled national or international air service.

Airman's Information Manual
A publication containing basic flight information and ATC procedures designed primarily as a pilot's instructional manual for use in the National Airspace System of the United States.

air mass
An extensive body of air having the same properties of temperature and moisture in a horizontal plane.

air mass classification
A system used to identify and to characterise the different air masses according to a basic scheme. The system most commonly used classifies air masses primarily according to the thermal properties of their source regions: 'tropical' (T), 'polar' (P), and 'Arctic' or 'Antarctic' (A). They are further classified according to moisture characteristics, as 'continental' (c) and 'maritime' (m).

Airmen advisory [US]
Information with regard to airport conditions or other 'useful to know' data, furnished to a pilot on request, by voice, teletype or by other means while en-route, or prior to departure or landing.

Airmet [UK]
A pre-recorded weather forecast available from telephone or telex services. It covers regions of the United Kingdom, and is read out at dictation speed. Forecasts are renewed routinely four times daily and become available about half an hour before the start of the period of validity.

Airmet (Airmen's meteorological information) [US]
In-flight weather advisories issued only to amend the area forecast concerning weather phenomena that are of operational interest to all aircraft, and potentially hazardous to aircraft having limited capability because of lack of equipment, instrumentation or pilot qualifications. Airmets concern weather of less severity than that covered by Sigmets of Convective sigmets. Airmets cover moderate icing, moderate turbulence, sustained winds of 30 knots or more at the surface, widespread areas of ceilings less than 1000 ft and/or visibility less than 3 miles and extensive mountain obscuration. See also CONVECTIVE SIGMET and SIGMET.

air miles per gallon
The air distance flown per gallon of fuel consumed, or air miles per pound/kg.

airmiss
An airmiss incident is when a pilot considers that a definite risk of collision exists between his and other aircraft during flight. The degree of collision risk is assessed as follows:

(A) Actual risk of collision. When it can definitely be established that there was a danger of collision.
(B) Possible risk of collision. When an actual risk of collision cannot be established but the aircraft concerned came into such proximity that the possibility of a collision occurred.
(C) No risk of collision. When it can be positively determined that no risk of collision existed at any time.
(D) Risk not determined. When there is either insufficient information or a conflict of evidence such that an assessment of collision risk cannot be made.

Air Navigation Commission [ICAO]
The principle body within ICAO concerned with the development of standards and recommended practices. The commission is composed of 15 persons who have suitable qualifications and experience in the science and practice of aeronautics. Its members nominated by the contracting States are appointed by the ICAO Council. The Commission reports to the Council and is responsible for the examination, co-ordination and planning of all ICAO's work in the air navigation field.

Air Navigation Council
The governing body of ICAO composed of 33 representatives from the Contracting States. It is a permanent body elected by, and responsible, to the ICAO Assembly. One of the major duties of the Council is to approve and adopt international standards and recommended practices and to incorporate these as Annexes to the Convention and International Civil Aviation. The Council may act as arbiter between Member States on matters concerning aviation and implementation of the Convention; it may investigate any situation which presents avoidable obstacles to the development of international air navigation and, in general, .it may take whatever steps are necessary to maintain the safety and regularity of operation of international air transport.

air navigation facility
Any facility used in, available for use in, or designed for use in aid of air navigation, including landing areas, lights, any apparatus or equipment for disseminating weather information, for signalling, for radio-directional finding, or for radio or other electrical communication, and any other structure or mechanism having a similar purpose for guiding or controlling flight in the air or the landing and takeoff of aircraft. See also NAVIGATIONAL AID.

Air Navigation Order
The UK statutory instrument defining the Articles and Schedules concerned with the enactment of air navigation issues.

Air operator's certificate [UK]
A certificate granted by the CAA and required by aircraft operators who engage in public transport operations.

airphone
Airline passenger telephone system.

airplane (US)
An engine-driven fixed wing aircraft that is heavier than air, and is supported in flight by the dynamic reaction of the air against its wings.

air plot
The position plotted from true heading and air speed, as though there were no wind. By way of contrast a position determined after taking into account the effect of wind (that is, from track and ground speed) is a 'ground plot'.

airport
An area of land or water that is used, or intended to be used, for the landing and takeoff of aircraft, and including its buildings and facilities, if any.

airport advisory area [US]
The area within 10 miles of an airport without a control tower, or where the tower is not in operation and on which a flight service station is located. See also AIRPORT ADVISORY SERVICE.

airport advisory service [US]
A service provided by a flight service station at an airport not served by a control tower. This service consists of providing information to arriving and departing aircraft concerning wind direction and speed, favoured runway, altimeter setting, pertinent known traffic, pertinent known field conditions, airport taxi routes and traffic patterns, and authorised instrument approach procedures.

airport capacity
Airport capacity is measured from the combined result of the performance of one or more components of the three main airport subsystems: the terminal, apron, and aircraft movements per hour. The terminal capacity is the number of passengers per hour which depends upon the movement rate through various junctions in the terminal building, e.g. security, immigration, customs, etc. The apron capacity is the number of aircraft handled per hour which depends upon the number of parking stands and the capability of the ground handling agency to service the associated aircraft. The aircraft movement capacity is the number of movements per hour that the combined ATC services, runways and taxiways, etc can support.

airport control tower
A facility providing airport traffic control service for aircraft operations on and in the vicinity of an airport.

airport elevation
The highest point of an airport's usable runways, measured in feet from mean sea level. See also TOUCH DOWN ZONE ELEVATION.

Airport/Facility Directory [US]
A publication designed primarily as a pilot's operational manual containing all airports, seaplane bases and heliports open to the public; including communications data, navigational facilities and certain special notices and procedures. This publication is issued in seven volumes according to geographical area.

airport information desk [US]
An airport unmanned facility designed for pilot self service briefing, flight planning and filing of flight plans.

airport lighting
Various lighting aids that may be installed on an airport. Types of airport lighting include:

(1) Approach lighting systems. Airport lighting facilities which provide visual guidance to landing aircraft by radiating light beams in a horizontal or vertical directional pattern by which the pilot aligns the aircraft with the extended centreline of the runway and/or determines his vertical position on his final approach for landing.

Types of approach lighting systems.

(i) Approach lighting system with sequenced flashing lights.
(ii) PAPI   See PRECISION APPROACH PATH INDICATOR SYSTEM.
(iii) Runway alignment indicator lights (sequenced flashing lights that are installed only in combination with other light systems).
(iv) Sequenced flashing lead-in lights.

(2) Runway lighting. A number of different lighting systems are used on runways and taxiways, typical of which are:

(i) Runway centre lighting. Flush centreline lights spaced at intervals along the runway.
(ii) Runway edge lights. Lights having a prescribed angle of emission, used to define the lateral limits of a runway. Runway lights are uniformly spaced at intervals and the intensity may be controlled or preset.
(iii) Runway end lights. These consist of at least six unidirectional lights showing red in the direction of the runway.
(iv) Runway guard lights. These are provided on taxiways at access points to runways and used to prevent inadvertent incursion by aircraft and vehicles.
(v) Threshold lights. Fixed green lights arranged symmetrically left and right of the runway centreline, identifying the runway threshold.

(3) Taxiway lighting.

(i) Taxi-holding position lights. These should be provided at a precision approach Category II or III taxi-holding position. Where provided they will be located at each side of a taxi-holding position as close as possible to the taxiway edge and consist of two alternately illuminated yellow lights.
(ii) Taxiway centreline lights. Fixed lights along the centreline of the taxiway.
(iii) Taxiway edge lights. Normally provided on a holding bay, apron, etc, intended for use at night and on a taxiway not provided with taxiway centreline lights and intended for use at night. Usually blue in colour.
(iv) Taxiway stop bars. Normally provided at a taxiway intersection or taxiway-holding position when it is desired to provide traffic control by visual means. Stop bar lights are red in colour.

airport marking aids
Markings used on runway and taxiway surfaces to identify a specific runway, a runway threshold, a centreline, a hold-line, etc. A runway should be marked in accordance with its present usage such as:

(1) Visual
(2) Non-precision instrument.
(3) Precision instrument.

airport rotating beacon
A visual navaid operated at many airports. At civil airports alternating white and green flashes indicate the location of the airport.

airport scheduling
Co-ordination of arrivals and departures of planned flight operations at a given airport or group of airports, normally done by operators concerned in co-ordination with the responsible authorities.

airport surface detection equipment
Radar equipment specifically designed to detect all principal features on the surface of an airport, including aircraft and vehicular traffic, and to present the entire image on a radar indicator console in the control tower. Used to augment visual observation by tower personnel of aircraft and/or vehicular movements on runways and taxiways.

airport surveillance radar
Approach control radar used to detect and display an aircraft's position in the terminal area. ASR provides range and azimuth information but does not provide elevation data.

airport traffic area [US]
Unless otherwise specifically designated in FAR Part 93, that airspace within a horizontal radius of five statute miles from the geographical centre of any airport at which a control tower is operating, extending from the surface up to, but not including, an altitude of 3000 ft above the elevation of the airport. Unless otherwise authorised or required by ATC, no person may operate an aircraft within an airport traffic area except for the purpose of landing at, or taking off from, an airport within that area. ATC authorisation may be given as individual approval of specific operations or may be contained in written agreements between airport users and the tower concerned.

airport traffic control service
A service provided by a control tower for aircraft operating on the movement area and in the vicinity of an airport. See also MOVEMENT AREA.

air-report
A report from an aircraft in flight prepared in conformity with requirements for position, and operational and/or meteorological reporting.

air route facilities
Facilities provided to permit safe operation of aircraft along an air route, including visual and radio navigation aids for approach and landing at aerodromes, and

communication services, meteorological services, and air traffic services and facilities.

air route surveillance radar
Air route traffic control centre (ARTCC) radar, used primarily to detect and display an aircraft's position while en-route between terminal areas. It enables controllers to provide radar air traffic control service when aircraft are within radar coverage. In some instances it may enable an ARTCC to provide terminal radar services similar to, but usually more limited than, those provided by a radar approach control.

air route traffic control centre [US]
A facility established to provide air traffic control service to aircraft operating on IFR flight plans within controlled airspace and principally during the en-route phase of flight. When equipment capabilities and controller workload permit, certain advisory/assistance services may be provided to VFR aircraft. See EN-ROUTE AIR TRAFFIC CONTROL SERVICE.

air–sea rescue
Aircraft operation for the purpose of saving life at sea.

air service
Any scheduled air service performed by aircraft for the public transport of passengers, mail or cargo.

airship
An engine-driven, lighter than air aircraft, that can be steered.

air side
The movement area of an airport, adjacent terrain, and buildings or portions thereof, access to which is controlled.

airspace and traffic management
A generic term covering the co-operative activities on the part of authorities concerned with planning for the organisation of the most effective use of the airspace and the handling of the air traffic flows within their area of responsibility.

Airspace and Traffic Management Group
A sub-group of the European Air Navigation Planning Group (ICAO).

airspace management
A generic term covering the co-operative activities on the part of authorities concerned with the planning for the organisation of the most effective exploitation of the airspace in accordance with the legitimate requirements of the various airspace users.

airspace reservation
A defined volume of airspace normally under the jurisdiction of one aviation authority and temporarily reserved, by common agreement, for exclusive use by another aviation authority.

airspace volume concept
A concept of controlled airspace organisation which allows an aircraft operator complete freedom to manoeuvre within a designated airspace.

air speed
The velocity of an aircraft with respect to the air. Air speed may be indicated (IAS), as read directly from the instrument; calibrated (CAS), the IAS corrected for instrumental and installation errors; or true (TAS), the CAS corrected for altitude and temperature and (when necessary) for compressibility. Equivalant air speed (EAS) is CAS corrected for compressibility.
Note: In the UK the term RAS is used instead of CAS.

airspeed indicator
An instrument used for measuring the speed of an aircraft through the air.

airstairs
A short stairway for use in embarking and disembarking from an aircraft. Fitted in unit form so that it can be folded up as part of a door or stowed on board.

airstart
The starting of an aircraft engine while the aircraft is airborne, preceded by engine shutdown during training flights or by actual engine failure.

airstrip
A landing strip normally of small dimensions. Used in many general aviation operations.

air taxi
(1) Used to describe a helicopter/VTOL aircraft movement conducted above the surface but normally not above 100 ft agl. The aircraft may proceed either via hover taxi or flight at speeds more that 20 knots. The pilot is solely responsible for selecting a safe airspeed/altitude, for the operation being conducted.
(2) An aircraft, generally below 5700 kg being used for ad hoc passenger carrying flights for remuneration.

air-to-ground communication
One-way communication from aircraft to stations or locations on the surface of the earth.

Air traffic
Aircraft operating in the air or on an aerodrome surface, exclusive of loading ramps and parking areas.

air traffic [ICAO]
All aircraft in flight or operating on the manoeuvring area of an aerodrome.

air traffic advisory service
A service provided within advisory airspace to ensure separation, in so far is as possible, between aircraft.

air traffic clearance
An authorisation by air traffic control, for the purpose of preventing collisions between known aircraft, for an aircraft to proceed under specific traffic conditions within controlled airspace. See ATC INSTRUCTIONS.

air traffic control
A service operated by an appropriate authority to promote the safe, orderly and expeditious flow of air traffic.

air traffic control centre
A unit providing many elements of air traffic service for a specific area.

air traffic control clearance
Authorisation for an aircraft to proceed under conditions specified by an air traffic control unit.
Note 1: For convenience, the term 'air traffic control clearance' is frequently abbreviated to 'clearance' when used in appropriate contexts.
Note 2: The abbreviated term 'clearance' may be prefixed by the words 'taxi', 'takeoff', 'departure', 'en-route', 'approach' or 'landing' to indicate the particular portion of flight to which the air traffic control clearance relates.

air traffic control officer
A person authorised to provide air traffic control service. See AIR TRAFFIC CONTROL SERVICE, FLIGHT SERVICE STATION.

air traffic control radar beacon system
A system of radar beacons which trigger coded radar responses from airborne transponders, providing identification, position and (sometimes) altitude of aircraft in flight, for traffic control.

air traffic control service [ICAO]
A service provided for the purpose of:

(1) Preventing collisions
  (i) between aircraft, and
  (ii) on the manoeuvring area between aircraft and obstructions.

(2) Expediting and maintaining an orderly flow of air traffic.

air traffic control systems command centre [US]
An air traffic service facility consisting of four operational units.

(1) Central flow control function. Responsible for co-ordination and approval of all major inter-centre flow control restrictions on a system basis, in order to obtain maximum use of the airspace. See also FUEL ADVISORY DEPARTURE and QUOTA FLOW CONTROL.
(2) Central altitude reservation function. Responsible for co-ordinating, planning and approving special user requirements under the altitude reservation concept. See also ALTITUDE RESERVATION.
(3) Airport reservation office. Responsible for approving IFR flights at designated high density traffic airports (John F. Kennedy, La Guardia, O'Hare and Washington National) during specified hours. (Refer to *US Airport Facility Directory*; FAR part 93).
(4) ATC contingency command post. A facility that enables the FAA to manage the ATC system when significant portions of the system's capabilities have been lost or are threatened.

air traffic control tower
An ATC operations facility situated at an aerodrome using air–ground communica-

tions, e.g. radio, visual signals and other means to provide safe and efficient movement of aircraft.

air traffic control unit
A generic term meaning, variously, area control centre, approach control office or aerodrome control tower.

air traffic demand
(1) For airspace management purposes the amount of planned flight operations of all airspace users in a given time period and related to a given area, route or location.
(2) For ATC and ATFM purposes that portion of the total air traffic demand requiring the provision of air traffic control service.

air traffic flow management position
Working position established within an ACC to ensure co-ordination between the associated air traffic flow management unit and the ACC in question on matters concerning the provision of the air traffic flow management service.

air traffic flow management service
An air traffic service to ensure an optimum flow of air traffic to or through areas within which air traffic demand at times exceeds the available sustainable capacity of the ATC system.

air traffic flow management unit
An ATS unit charged with the provision of the air traffic flow management service within the area of responsibility of one or more area control centres.

air traffic management
Consists of a ground part and an air part, both needed to ensure the safe and efficient movement of aircraft. The execution of this calls for a close integration of the ground part and the air part through well defined procedures and interfaces. The functions of air traffic management include air traffic control, air traffic flow management and airspace management. Air traffic control (ATC) is the primary component of the management system. The main objectives of ATC are to prevent collisions between aircraft, and between aircraft and obstructions in the manoeuvring area, and to expedite and maintain an orderly flow of air traffic. Flow management is a necessary adjunct to the air traffic control service to assist ATC in the attainment of its objectives, and in achieving the fullest possible exploitation of available system capacity. The objective of flow management is to ensure an optimum flow of air traffic to or through areas within which the demand at times exceeds the available capacity of the ATC system. The objective of airspace management is to avoid permanent airspace segregation and to achieve time-sharing use of the airspace based on actual needs. In addition, the total system is dedicated to provide a service whereby operators may achieve their planned times of departure and arrival, with adherence to their proposed profiles with an absolute minimum of constraint.

air traffic service
A generic term meaning, variously, flight information service, alerting service, air traffic advisory service, air traffic control service, area control service, approach control service or aerodrome control service.

air traffic services reporting office
A unit established for the purpose of receiving reports concerning air traffic services and flight plans submitted before departure.
Note: An air traffic services reporting office may be established as a separate unit or combined with an existing unit, such as another air traffic services unit, or a unit of the aeronautical information service.

air traffic services unit
A generic term meaning, variously, air traffic control unit, flight information centre or air traffic services reporting office.

air transportation
Domestic, overseas, or foreign air transportation of passengers and/or the transportation of mail or other cargo by aircraft.

air transport undertaking
An undertaking whose business includes the carriage by air of passengers or cargo for hire or reward.

airway [ICAO]
A control area, or portion thereof, established in the form of a corridor equipped with radio navigational aids.

airway beacon [US]
Used to mark airway segments in remote mountain areas. The light flashes a morse code to identify the beacon site.

airworthiness approval certificate
A certificate issued by a responsible authority concerning items of aircraft equipment of foreign origin.

Airworthiness Approval Note [UK]
When an aircraft made by a foreign manufacturer, having a type certificate already issued in the manufacturer's country, is imported into the UK together with an export Certificate of Airworthiness, the UK airworthiness division of the CAA will, if satisfied that the aircraft meets the UK airworthiness requirements, issue an Airworthiness Approval Note. Normally one Airworthiness Approval Note will cover any more imported aircraft of the same model, identical in design, construction and performance.

Airworthiness directive
Notification from a national airworthiness authority requiring an immediate mandatory inspection or modification to an aircraft or its equipment.

Airworthiness notice
A notice published by airworthiness authorities for the purpose of circulatory information to all those concerned with the airworthiness of civil aircraft.

airworthy
Describes an aircraft which meets all regulations and requirements of the national airworthiness authority.

albedo
The ratio of the amount of electromagnetic radiation reflected by a body to the amount incident upon it, commonly expressed in percentages. In meteorology, normally used in reference to insolation (solar radiation); i.e., the albedo of wet sand is 9, meaning that about 9% of the incident insolation is reflected; albedoes of other surfaces range upward to 80–85 for fresh snow cover; average albedo for the earth and its atmosphere has been calculated to range from 35 to 43.

Aldis lamp   See LIGHT GUN.

Alerfa
Code word used to designate an alert phase.

alert area
An area which may contain a high volume of any unusual air activity. Pilots need to be especially alert in such areas, but flight is not otherwise restricted.

alerting post
A unit designated to receive information from the general public regarding aircraft in emergency and to forward the information to the associated rescue co-ordination centre.

alerting service
A service provided to notify appropriate organisations regarding aircraft in need of search and rescue aid, and to assist such organisations as required.

Alert notice [US]
A message sent by a flight service station or air route traffic control centre that requests an extensive communication search for overdue, unreported or missing aircraft.

alert phase
A situation wherein apprehension exists as to the safety of an aircraft and its occupants.

all-call
An intermode or Mode S interrogation that can elicit replies from more than one transponder.

all-call address
The transponder Mode S address consisting of 24 consecutive tones when used in Mode S broadcast interrogations (also known as broadcast address).

all-call reply
A transponder Mode S reply containing the 24-bit Mode S address of the transponder installation plus a capability report in its message field. The all-call reply provides the Mode S ground station with the information needed selectively to address the transponder in subsequent Mode S interrogations.

all-electric aeroplane
A future concept aeroplane without hydraulic or mechanical systems.

allowable deficiency
A missing, inoperative or imperfectly functioning item which does not invalidate the aircraft certificate of airworthiness.

alpha hinge (helicopter)
A common designation or term for the drag hinge on an articulated rotor blade. Gradual pick up of the rotor's momentum while the system is being engaged. Many helicopters are also able, via their clutch system, to disengage the rotor while the engine is running; this can be useful during short stop-overs.

alphanumeric characters (alphanumerics)
A collective term for letters and figures (digits).

alphanumeric display/data block
Letters and numerals used to show identification, altitude, beacon code and other information concerning a target on a radar display. See AUTOMATED RADAR TERMINAL SYSTEMS.

alternate aerodrome for takeoff
An aerodrome to which a flight may proceed when the weather conditions at the aerodrome of departure would preclude an immediate return for landing.

alternate aerodrome [ICAO]
An aerodrome specified in the flight plan to which a flight may proceed when it becomes inadvisable to land at the aerodrome of intended landing.
Note: An alternate aerodrome may be the aerodrome of departure.

alternate airport
An airport at which an aircraft may land if a landing at the intended airport becomes inadvisable.

alternating current
A current which flows from zero to a maximum in one direction, and then reverses through zero to a maximum in the opposite direction. The number of cycles per second is called the frequency.

alternating light
An intermittent light flashing in two or more alternate colours (normally white/green).

alternative means of communication
A means of communication provided with equal status with, and in addition to, the primary means.

alternative route(s)
Route(s) other than the preferred route(s) between given points or areas of origin and destination, constituting a route selected by aircraft operators for cases when the preferred route is not available or is subject to delay.

alternator
An electric generator which produces alternating current.

altimeter
An instrument designed to indicate the approximate altitude above mean sea level by measuring the static air pressure.

altimeter errors (pressure activated)
Mechanical errors are relatively small in a correctly functioning altimeter, and although installation error brought about when the static tube is not aligned directly into the airflow is not very large, it can be reduced even further by the use of static vents mounted flush with the fuselage. Whenever the air temperature varies from that specified in the requirements for the international standard atmosphere, an additional error will occur and the altimeter will either under-read or over-read. For example, an altimeter at sea level will be measuring the weight of a column of air above it. If the aircraft has now climbed to 5000 ft, the weight or pressure on the altimeter aneroid is less by that portion of the column of air below it. If, under these conditions, the temperature is the same as the ISA, then the altimeter will read 5000 ft. However, if the lower portion of the column is warmer than ISA, then the air will expand and the reduction of pressure will be smaller in this column of air and the altimeter will under-read. Conversely, if the temperature in the column of air is below that of the ISA, the altimeter will over-read. Temperature error is related to density error and given the pressure altitude and the air temperature at that altitude the error can be calculated by using a navigational computer.

altimeter settings
The purpose of an aircraft altimeter is to measure distance above the surface and this distance is referred to by the use of three different terms, altitude, height and flight level. Each term is defined in relation to a different datum as follows:

(1) Altitude. The vertical distance of a level, point or object measured from mean sea level and is related to a datum setting known as QNH.
(2) Height. The vertical distance of a level, point or object, measured from an aerodrome surface. This datum is known as the aerodrome QFE.
(3) Flight level. A level measured in relation to a constant atmospheric pressure datum known as the standard setting or QNE.

In each of the above cases a different millibar reference datum is used and, when obtained this reference datum is set on the sub-scale of the altimeter by means of the pressure setting knob. When the millibar setting is increased the altimeter reading will increase, and when the millibar setting is decreased the altimeter reading is decreased. An aircraft altimeter is essentially an aneroid barometer in which the scale is constructed to indicate increments of height or altitude in feet, rather than units of pressure. It must nevertheless be appreciated that its principle is based upon the measurement of atmospheric pressure. In other words, if an aircraft is maintaining an indicated altitude of 4000 ft, the pilot will be following the horizontal atmospheric pressure which gives this indication. Atmospheric pressure is measured in millibars (or hectopascals) or inches of mercury and lines of equal pressure are known as isobars. These can be depicted in plan form as shown on surface weather charts.

altimetric valve
A device which is sensitive to cabin pressure.

altitude

The distance above mean sea level; the angular distance of a body above the horizon; the vertical distance of a level, a point or an object, considered as a point measured from mean sea level (amsl). In instrument flight the altitude in feet is often referred to other datum planes, designated by appropriate abbreviations, as follows:

(1) AGL. The height of an aircraft above ground level, also known as absolute altitude; may be determined by radio altimeter, or by subtracting the elevation of the ground from the flight altitude above mean sea level.

(2) DH. The height in feet above mean sea level at which a decision must be made either to continue an instrument approach or to begin a missed approach.

(3) HAA. The height above the published airport elevation; it is determined as in AGL.

(4) HAT. The height above touchdown; that is, the height of the DH (decision height) or MDA (minimum descent altitude) above the highest runway elevation in the touchdown zone (first 3000 ft of runway). It is obtained by subtracting the ground elevation of the point of touchdown from the flight altitude above mean sea level.

(5) MCA. The minimum crossing altitude at which certain radio facilities or intersections must be crossed in specified directions of flight. If a normal climb, commenced immediately after passing a fix beyond which a higher MEA applies, would not insure adequate obstruction clearance, an MCA is specified. The MCA at certain points could be lower for a departing aircraft than for an en-route aircraft; the lower MCA would not be shown on the en-route chart but might be received in an IFR clearance.

(6) MDA. The minimum descent altitude. It is the lowest altitude above mean sea level to which descent is authorised on final approach when no electronic glide slope is available.

(7) MEA. The minimum en-route altitude between radio fixes. It is established to meet obstruction clearance requirements and to provide for adequate reception of radio navigational signals. MEA is sometimes different for flight in opposite directions due to rising or lowering terrain.

(8) MOCA. The minimum obstruction clearance altitude. It is the same as MEA, except that adequate reception of navigational signals is assured only within a limited distance of the VOR.

(9) MRA. The minimum reception altitude. It is the lowest altitude above sea level at which accurate determination of position can be made at a specified intersection, using the facilities that establish the intersection.

(10) MSA. The minimum sector altitude.

(11) SAA. The safe sector altitude.

See also CORRECTED ALTITUDE, DENSITY ALTITUDE, INDICATED ALTITUDE, PRESSURE ALTITUDE, RADAR ALTITUDE and TRUE ALTITUDE.

altitude clearance

A clearance given by air traffic control for flight at a specific altitude.

altitude datum

Horizontal level from which altitude is measured.

altitude hole
Loss of signal in radar due to round trip travel time being equal to the modulation period; of importance in Doppler navigation systems.

altitude readout
An aircraft's altitude, transmitted via the Mode C transponder feature, that is visually displayed in 100 ft increments on a radar scope having readout capability. See also ALPHA NUMERIC DISPLAY.

altitude reporting
Automatic coded transmission of altitude from aircraft to ATC in an SSR system.

altitude request (transponder)
A surveillance of Comm-A interrogation that elicits the altitude code in the associated reply.

altitude reservation [US]
Airspace utilisation under prescribed conditions normally employed for the mass movement of aircraft or other special user requirements that cannot otherwise be accomplished.

altitude restriction
An altitude or altitudes stated in the order flown, which are to be maintained until reaching a specific point or time. Altitude restrictions may be issued by ATC due to traffic, terrain or other airspace considerations.

altitude restrictions are cancelled
Adherence to previously imposed altitude restrictions is no longer required during a climb or descent.

altitude sickness
Nausea, etc. caused by the body being exposed to less than normal atmospheric pressure.

altocumulus
White or grey layers or patches of cloud, often with a waved appearance; cloud elements appear as rounded mases or rolls; composed mostly of liquid water droplets which may be super-cooled; may contain ice crystals at sub-freezing temperatures.

altocumulus castellanus
A species of middle cloud of which at least a fraction of the upper part presents some vertically, developed, cumuliform protuberances (some of which are taller than they are wide, as castles) and which give the cloud a crenelated or turreted appearance especially evident when seen from the side; elements usually have a common base arranged in lines. This cloud indicates instability and turbulence at the altitudes of occurrence.

altostratus
A form of middle cloud.

aluminium (Al)
Metal element, atomic weight 26.9815, atomic number 13, relative density 2.7, melting point 659.7°C. Good conductor of electricity.

amatol
Explosive made from TNT (20%) and ammonium nitrate (80%).

ambient
Surrounding or pertaining to the immediate environment.

amended clearance
A clearance issued by air traffic control amending a previous ATC clearance, usually in response to a request from the pilot, or initiated to avoid conflict with other aircraft or severe weather.

ammeter
Instrument for measuring amperes of electric current.

ampere
The unit of electric current, approximately 6×10^{18} electrons per second (1 coulomb per second).

ampere hour
The flow of 1 ampere for 1 hour (3600 Coulombs).

ampere turns
Measuring the magnetomotive force by the current in a coil × the number of turns in the winding.

amphibian
An aircraft designed to operate from land or water.

amplifier
Device which receives weak input signals and uses electrical power to increase their strength.

amplitude
The maximum variation, positive and negative from zero (or equilibrium). For waves, the square of the amplitude indicates the quantity of energy carried.

amplitude modulation
A method of sending voice data by radio transmission.

anabatic
A wind which flows up sloping ground when the ground surface has an appreciable slope which becomes heated by the sun; the air in contact with the slope becomes warmer than the air at the same level over the lower land, and thus is lighter and tends to ascend.

analog
A quantity or signal which varies continuously and represents some other continuously varying quantity; hence an analog circuit which processes such signals, an analog computer which performs arithmetic operations on such signals.

analogue   See ANALOG.

anchor nut
A nut which is positively secured by means of a screwed or bolted plate at the base.

anemograph
An instrument designed to show a permanent record of wind velocity.

anemometer
Instrument used for measuring wind velocity.

aneroid
'Without liquid'. For example, the altimeter is a form of aneroid barometer in which changes in air pressure compress or expand capsules containing a partial vacuum.

aneroid barometer
A barometer which operates on the principle of having changing atmospheric pressure bend a metallic surface which, in turn, moves a pointer across a scale graduated in units of pressure.

angel
In radar meteorology, an echo caused by physical phenomena not discernible to the eye; angels have been observed when abnormally strong temperature and/or moisture gradients were known to exist; sometimes attributed to insects or birds flying in the radar beam.

angle of approach
The angle relative to the horizontal which is adopted by the pilot to initiate and maintain the approach path to a landing.

angle of attack
The angle at which the chord of an aircraft wing meets the relative airflow.

angle of incidence
The angle between the chord line of the main plane and the horizontal when the aircraft is in the rigging position.

Angström
A unit of length $= 10^{-10}$ m.

angular displacement sensitivity
The ratio of measured DDM to the corresponding angular displacement from the appropriate reference line.

anhedral
The angle of fall of an aerofoil measured from root to tip.

anion
A negatively charged ion attracted to the anode.

annual variation
The yearly variation of the position of magnetic north.

annunciator
An indicator visible to the pilot which shows the synchonisation of gyro compass with the correct magnetic direction.

anode
The positive pole or electrode.

anomalous propagation
In radar meteorology, the greater than normal bending of the radar beam such that echoes are received from ground targets at distances greater than normal ground clutter. Sometimes called anoprop.

antenna
Aerial.

anti-balance tab
A control surface tab arranged to deflect in the same direction as the main control in order to damp the main control movement.

anticipated operating conditions
Those conditions which are known from experience or which can be reasonably envisaged to occur during the operational life of the aircraft, taking into account the operations for which the aircraft is made eligible, the conditions so considerd being relative to the meteorological state of the atmosphere, to the configuration of terrain, to the functioning of the aircraft, to the efficiency of personnel and to all the factors affecting safety in flight. Anticipated operating conditions do not include:

(1) those extremes which can be effectively avoided by means of operating procedures; and
(2) those extremes which occur so infrequently that to require the standards to be met in such extremes would give a higher level of airworthiness than experience has shown to be necessary and practical.

anti-corrosion
A surface coating applied to aircraft structures to reduce corrosion of the material.

anticyclone
An area of high atmospheric pressure which has a closed circulation that is anticyclonic, i.e. as viewed from above the circulation is clockwise in the northern hemisphere, counter clockwise in the southern hemisphere, and undefined at the equator.

anti-icing
Measures used to prevent ice formation on aircraft.

anti-knock rating
A measure used to indicate the resistance to detonation of fuel.

anvil cloud
Popular name given to the top portion of a cumulonimbus cloud having an anvil-like form.

aperiodic
Damping which reduces or prevents dynamic oscillation.

appliance
Any instrument, mechanism, equipment, part, apparatus, appurtenance or accessory, including communications equipment, that is used or intended to be used in operating or controlling an aircraft in flight, is installed in or attached to the aircraft, and is not part of the airframe, engine or propeller.

approach clearance
Authorisation by ATC for a pilot to conduct an instrument approach. The type of instrument approach and other pertinent information are provided in the approach clearance when required. See also INSTRUMENT APPROACH PROCEDURE and CLEARED FOR APPROACH.

approach control
A unit established to provide air traffic control service to controlled flights arriving at, or departing from, one or more aerodromes.

approach control facility
A terminal air traffic control facility providing approach control service in a terminal area. See also APPROACH CONTROL SERVICE.

approach control office
A unit established to provide air traffic control service to controlled flights arriving at, or departing from, one or more aerodromes.

approach control service
An air traffic control service provided by an approach control facility for arriving and departing VFR/IFR aircraft and, on occasion, en-route aircraft. At some airports not served by an approach control facility, the ARTCC provides limited approach control service.

approach control service [ICAO]
Air traffic service for arriving or departing controlled flights.

approach fix
The fix from which final IFR approach to an airport is executed.

approach funnel
A specified airspace around a nominal approach path within which an aircraft approaching to land is considered to be making a normal approach.

approach gate
The point on the final approach course one mile from the approach fix on the side away from the airport, or 5 miles from the landing threshold, whichever is farther from the landing threshold.

approach lighting system
A series of lights providing aid in a visual approach to a runway after breakthrough on an instrument approach. Available systems include: high and medium intensity lights; runway end identification lights (REIL); visual approach slope indicator lights (VASI) and others.

approach phase
The operating phase defined by the time during which the engine(s) is(are) operated in the approach operating mode.

approach plate
A document published to give details of specific approach aids and the required procedures.

approach sequence [ICAO]
The order in which two or more aircraft are cleared to approach to land at the aerodrome.

approach speed
The recommended speed contained in aircraft manuals used by pilots when making an approach to landing. This speed will vary for different segments of an approach as well as for aircraft weight and configuration.

appropriate airworthiness requirement [ICAO]
The comprehensive and detailed airworthiness codes established by a contracting State for the class of aircraft under consideration.

appropriate ATS authority
The relevant authority designated by the State responsible for providing air traffic services in the airspace concerned.

appropriate authority
(1) Regarding flight over the high seas: the relevant authority of the State of registry.
(2) Regarding flight other than over the high seas: the relevant authority of the State having sovereignty over the territory being overflown.

approved
Approved by the Aviation Authority, unless used with reference to another person.

approved maintenance organisation [ICAO]
An organisation approved by a contracting State to perform inspection, overhaul, maintenance, repair and/or modification of aircraft or parts thereof, and operating under supervision approved by that State.

approved training [ICAO]
Training carried out under special curricula and supervision approved by a contracting State.

apron [ICAO]
A defined area, on a land aerodrome, intended to accommodate aircraft for purposes of loading or unloading passengers or cargo, refuelling, parking or maintenance.

apron management service
A service provided to regulate the activities and the movement of aircraft and vehicles on an aerodrome apron.

apron
A defined area, on a land airport, intended to accommodate aircraft for purposes of loading or unloading passengers or cargo, refuelling, parking or maintenance.

arc
The track of an aircraft over the ground when flying at a constant distance from a navaid by reference to DME.

Arctic air
An air mass with characteristics developed mostly in winter over arctic surfaces of ice and snow. Arctic air extends to great heights, and the surface temperatures are basically, but not always, lower than those of polar air.

Arctic front
The surface of discontinuity between very cold (Arctic) air flowing directly from the Arctic region and another less cold and consequently less dense air mass.

area control centre
A unit established to provide air traffic control service to controlled flights in control areas under its jurisdiction.

area forecast
A weather forecast for a specific geographic area.

area minimum altitude
The lowest altitude to be used under instrument meteorological conditions (IMC) which will provide a minimum vertical clearance of 300 m (1000 ft) or in designated mountainous terrain 600 m (2000 ft) above all obstacles located in the area specified, rounded up to nearest (next higher) 30 m (100 ft).

area navigation
A method of navigation that permits aircraft operations on any desired course within the coverage of station-referenced navigation signals, or within the limits of self contained system capability.

(1) Area navigation low route (US). An area navigation route within the airspace extending upwards from 1200 ft above the surface of the earth to, but not including, 18 000 ft amsl.
(2) Area navigation high route (US). An area navigation route within the airspace extending upwards from and including 18 000 ft amsl to flight level 450.
(3) Random area navigation (RNAV) routes/random routes. Direct routes, based on area navigation capability, between waypoints defined in terms of degree/distance fixes or offset from published, or established routes/airways at specified distances and directions.
(4) RNAV waypoint. A predetermined geographical position used for route or instrument approach definition or progress reporting purposes.

area of coverage [WAFS]
A geographical area for which a regional area forecast centre supplies forecasts for flights departing from aerodromes in its service area.

area of responsibility [WAFS]
A geographical area for which a regional area forecast centre prepares significant weather forecasts.

armature
The rotating coils of an electric motor or dynamo.

army aviation flight information bulletin [US]
A bulletin that provides air operation data covering Army, National Guard and Army Reserve aviation activities.

arresting system
A safety device consisting of two major components, namely engaging or catching devices and energy absorption devices for the purpose of arresting both tail hook and/or non-tail hook equipped aircraft. It is used to prevent aircraft over-running runways when the aircraft cannot be stopped after landing or during aborted takeoff. Arresting systems have various names, e.g. arresting gear, hook, device, wire, barrier cable. See also ABORT.

arrival routes
Routes identified in an instrument approach procedure by which aircraft may proceed from the en-route phase of flight to an initial approach fix.

arrival time
The time an aircraft touches down on arrival.

articulated rotor (helicopter)
A rotor system which has its rotor blades hinged, so as to allow both flapping and lagging movements.

artificial horizon
A gyro-stablised horizon line fitted to an aircraft attitude indicator instrument.

A-scope
A type of radar display in which range is indicated by displacement along the trace and received signals cause a deflection at right angles to the trace.

aspect ratio
The span of an aerofoil divided by its average chord.

assignment
Distribution (of frequencies) to stations, distribution (of SSR codes) to aircraft.

associated identity
The identification of co-located VOR and DME beacons by synchronised transmission of the same morse code characters from each beacon.

Association of British Aviation Consultants [UK]
Objectives: An association of professional aviation consultants from varying disciplines, not agents or brokers, adhering to a professional code of conduct.

astronomical twilight See TWILIGHT.

asymmetric flight
A flight condition in which one engine, displaced from the aircraft's centre line, is inoperative.

ATC advises
Used to prefix a message of non-control information when it is relayed to an aircraft by other than an air traffic controller. See also ADVISORY SERVICE.

ATC assigned airspace [US]
Airspace of defined vertical/lateral limits, assigned by ATC, for the purpose of providing air traffic segregation between the specified activities being conducted within the assigned airspace and other IFR air traffic. See also ALERT AREA and MILITARY TRAINING AREA.

ATC clearance   See AIR TRAFFIC CLEARANCE

ATC clears
Used to prefix an ATC clearance when it is relayed to an aircraft by other than an air traffic controller.

ATC instructions
Directives issued by air traffic control for the purpose of requiring a pilot to take specific actions; e.g. 'turn left heading two five zero', 'go around', 'clear the runway'.

ATC requests
Used to prefix an ATC request when it is relayed to an aircraft by other than an air traffic controller.

atmosphere
(1) Gases surrounding a planet.
(2) Unit of pressure that will support a column of mercury 76 cm (29.92 inches) high at 0°C, amsl at latitude 45°.

atmospheric pressure
The pressure exerted by the atmosphere as a consequence of gravitational attraction exerted upon the 'column' of air lying directly above the point in question. Also called barometric pressure.

atmospherics
Crackling heard in radio receivers caused by natural electrical disturbances in the atmosphere.

atom
The smallest neutral structure that can participate in a chemical reaction.

atomic number
The number of protons in the nucleus of an element.

ATS direct speech circuit
An aeronautical fixed service (AFS) telephone circuit, for direct exchange of information between air traffic service (ATS) units.

ATS route
A specified route designed for channelling the flow of traffic as necessary for the provision of air traffic services.
Note: The term ATS route is used to mean variously: airway, advisory route, controlled or uncontrolled route, arrival or departure route, etc.

attenuation
In radar meteorology, any process which reduces power density in radar signals.
(1) Precipitation attenuation. Reduction of power density because of absorption or reflection of energy by precipitation.
(2) Range attenuation. Reduction of radar power density because of distance from the antenna. It occurs in the outgoing beam at a rate proportional to $1/range^2$. The return signal is also attenuated at the same rate.

attitude director indicator
An instrument which demands attitude changes which, if executed, causes the aircraft to fly a path determined by radio or other sensors.

audibility
The range of frequencies of sound waves that can be heard by humans: 30–20 000 Hz.

audio frequency
Generally, frequencies between 20–25 000 Hz.

audio integrating system
The electronic interface between crew members and audio sources.

aurora
A luminous radiant emission over middle and high latitudes confined to the thin air of high altitudes and centred over the earth's magnetic poles. Called *aurora borealis* (northern lights) or *aurora australis*, according to its occurrence in the northern or southern hemisphere respectively.

authorised agent
A responsible person who represents an operator and who is authorised by, or on behalf of, such operator to act on all formalities connected with the entry and clearance of the operator's aircraft, crew, passengers, cargo, mail, baggage or stores.

autokinesis
The apparent movement of a stationary light. A condition which can be experienced if, during darkness, a stationary light is stared at for many seconds.

automated radar terminal system [US]
The generic term for the ultimate in functional capability afforded by several automation systems. Each differs in functional capabilities and equipment. ARTS plus a suffix Roman numeral denotes a specific system. A following letter indicates a major modification to that system. In general, ARTS displays for the terminal controller aircraft identification, flight plan data, other flight associated information, e.g. altitude and speed, and aircraft position symbols in conjunction with the

radar presentation. Normal radar co-exists with alphanumeric display. In addition to enhancing visualisation of the air traffic situation, ARTS facilitates intra/inter-facility transfer and co-ordination of flight information. These capabilities are made possible by specially designed computers and subsystems tailored to the radar and communication equipment and operational requirements of each automated facility. Modular design permits adoption of improvements in computer software and electronic technologies as they become available, while retaining the characteristics unique to each system.

(1) ARTS II. A programmable, non-tracking, computer-aided display subsystem capable of modular expansion. ARTS II systems provide a level of automated air traffic control capability at terminals having low to medium activity. Flight identification and altitude may be associated with the display of secondary radar targets. Also, flight plan information may be exchanged between the terminal and ARTCC.

(2) ARTS III. The beacon tracking level (BTL) of the modular programmable automated radar terminal system in use at medium to high activity terminals. ARTS III detects, tracks and predicts secondary radar-derived aircraft targets. These are displayed by means of computer-generated symbols and alphanumeric characters depicting flight identification, aircraft altitude, ground speed and flight plan data. Although ARTS III does not track primary targets, it displays the coincident with the secondary radar as well as the symbols and alphanumerics. The system has the capability of communicating with ARTCCs and other ARTS III facilities.

(3) ARTS IIIA. The radar tracking and beacon tracking level (RT and BTL) of the modular, programmable, automated radar terminal system. ARTS IIIA detects, tracks and predicts primary as well as secondary radar-derived aircraft targets. An enhancement of the ARTS III, this more sophisticated computer-driven system will eventually replace the ARTS IA system and upgrade about half of the existing ARTS III systems in the United States. The enhanced system will provide improved tracking, continuous data recording and fail-safe capabilities.

automatic altitude report   See ALTITUDE READOUT.

Automatic altitude reporting
That function of a transponder that responds to Mode C interrogations by transmitting the aircraft's altitude in 100 ft increments.

automatic carrier landing system
US Navy final approach equipment consisting of precision tracking radar coupled to a computer data link to provide continuous information to the aircraft, monitoring capability to the pilot and a backup approach system.

automatic direction finder
An aircraft radio navigation system (ADF) that senses and indicates the direction to an LF/MF nondirectional radio beacon (NDB) ground transmitter. Direction is indicated to the pilot as a magnetic bearing or as a relative bearing to the longitudinal axis of the aircraft depending on the type of indicator installed in the aircraft. In certain applications, such as military, ADF operations may be based on

airborne and ground transmitters in the VHF/UHF frequency spectrum. See also BEARING.

automatic frequency control
An error-correcting circuit which may be fitted to receiving equipment to compensate for any frequency variation in either the receiver circuits or the received signal.

automatic terminal information service [ICAO]
The provision of current, routine information to arriving and departing aircraft by means of continuous and repetitive broadcasts throughout the day, or a specified portion of the day.

automation
Using computerised, electronic controls for electrical or mechanical processes.

autorotation
A rotorcraft flight condition in which the lifting rotor is driven entirely by action of the air when the rotorcraft is in motion.

(1) Autorative landing/touchdown autorotation. Used by pilot to indicate that he will be landing without applying power to the rotor.
(2) Low level autorotation. Commences at an altitude well below the traffic pattern, usually below 100 ft agl and is used primarily for tactical military training.
(3) 180° autorotation. Initiated from a downwind heading and commenced well inside the normal traffic pattern. Go-around may not be possible during the later part of this manoeuvre.

auxiliary power unit.
A self-contained power unit on an aircraft providing electrical/pneumatic power to aircraft systems during ground operations.

auxiliary rotor
A rotor that serves either to counteract the effect of the main rotor torque on a rotorcraft or to manoeuvre the rotorcraft about one or more of its three principle axes.

avgas
Aviation gasoline.

aviation weather service [US]
A service provided by the National Weather Service (NWS) and the FAA that collects and disseminates pertinent weather information for pilots, aircraft operators and ATC. Available aviation weather reports and forecasts are displayed at each NWS office and each FSS. See also EN-ROUTE FLIGHT ADVISORY SERVICE, TRANSCRIBED WEATHER BROADCAST and PILOTS' AUTOMATIC TELEPHONE WEATHER ANSWERING SERVICE.

avionics
Aeronautical electronic equipment intended for use in flight.

Avogadro's law (hypothesis)
At the same temperature and pressure, equal volumes of all gases contain equal numbers of molecules.

Avogadro's number
The number of atoms or molecules in one mole of a substance: 6.02252×10^{23} mole^{-1}

avoirdupois
The original British system of weights: ounces (oz), pounds (lb), stones (st), etc.

axis
An imaginary line about which something rotates.

axis of no flapping
On a helicopter's rotor system, an axis with respect to which there is no first harmonic flapping of the rotor blades. Sometimes referred to as the 'no flapping axis'.

azimuth   See DIRECTION.

azimuth control
Cyclic pitch, rotorcraft.

azimuth error
Error in a radar bearing due to horizontal refraction.

B

back course sector [ILS]
The course sector which is situated on the opposite side of the localiser from the runway.

backfire
The premature ignition of fuel mixture in an engine cylinder causing a flame to travel through the still open inlet valve and along the induction manifold.

backing
Shifting of the wind in an anti-clockwise direction with respect to either space or time: opposite of veering. Commonly used by meteorologists to refer to a cyclonic shift (anti-clockwise in the northern hemisphere and clockwise in the southern hemisphere).

background luminance monitor
Used in conjunction with transmissometers in IRVR systems.

backlash
Play in driven mechanical linkage such as scanner drives.

backscatter
Pertaining to radar, the energy reflected or scattered by a target; an echo.

backtrack
Taxying along the runway in the opposite direction to the takeoff or landing direction.

baggage
Personal property of passengers or crew, carried on an aircraft by agreement with the operator.

bail out
Jump from an aircraft in flight for the purpose of parachuting to earth.

balance aerodynamic
A method of reducing the force required by a pilot when operating a control surface.

balance tab
Small surface hinged to the trailing edge of a control surface. Assists the pilot to move the control surface.

ballistic missile
A ground-to-ground missile that is aimed and propelled only at the beginning of its flight.

ballistics
The study of the flight of bullets, shells, rockets, etc.

ballonet
A flexible, gas-tight section inside the main envelope of an airship.

balloon
A non-power-driven aircraft constructed from a flexible non-porous bag which is inflated with helium or by heated air. Usually equipped with a suspended gondola or basket.

bank angle
The angle of the aircraft wings measured from the horizontal during a turn.

banner cloud
A banner-like cloud streaming off from a mountain peak. Also called cloud banner.

bar
Unit of pressure which is used in meteorology, one bar = 1000 millibars = $10^5 Nm^{-2}$. See also ISOBARS.

barograph
Meteorological instrument which records atmospheric pressure throughout the day (24 hours) on a revolving drum.

barometer
Meteorological instrument for measuring atmospheric pressure either by means of a column of mercury or a stack of capsules containing partial vacuums.

barometric pressure   See ATMOSPHERIC PRESSURE.

barometric tendency
The changes of atmospheric pressure within a specific time (usually three hours) before the observation.

barrette
Three or more aeronautical ground lights closely spaced in a transverse line so that from a distance they appear as a short bar of light.

base
The integral value of the number of symbols in a counting system; the central region of a bipolar transistor.

base leg   See TRAFFIC PATTERN.

base line
The line joining the principal point of a vertical photograph with the image position of the principal point of the next photograph in the run.

base turn
A turn executed by the aircraft during the initial approach between the end of the outbound track and the beginning of the intermediate or final approach track. The tracks are not reciprocal.

Note: Base turns may be designated as being made either in level flight or while descending, according to the circumstances of each individual procedure.

basic cover
Complete cover of an area with which more recent photographs may be compared.

basic data [ILS/MLS]
Data transmitted by the ground equipment that is associated directly with the operation of a landing guidance system, and advisory data on the MLS ground equipment performance level.

basic equipment [UK]
The unconsumable fluids, and the equipment which is common to all roles for which the operator intends to use the aircraft.

basic operational requirements
These, together with planning criteria and methods of application, form the basis for planning activities in the ICAO regions. They define the objectives for a uniform and coherent air navigation system that have to be in accordance with relevant ICAO worldwide provisions.

basic weight
The weight of the aircraft and all its basic equipment, plus that of the declared quantity of unusable fuel and oil. In turbine-engined aircraft and those which do not exceed a total weight authorised of more than 5700 kg, it may also include the weight of usable oil.

beacon
Non-directional beacon, marker beacon, airport rotating beacon, aeronautical beacon, airway beacon. An approved aeronautical light beacon; a radiobeacon; a radio crash locator beacon; radio personnel locator beacon; or survival radio beacon to help locate aircraft or personnel, etc. see also RADAR.

beam
Radiation travelling in one direction.

beam resolution   See RESOLUTION.

beam riding
Missiles following a beam of radar.

beam softening
Progressive reduction in gain of demand signal channel in landing systems.

beam transmission
The transmission of directional beams.

bearing
The angle, measured in a clockwise direction, between a reference line through the aircraft and a line joining the aircraft and the object to which bearing is being measured. The reference line may point to magnetic or true north or be in line with the aircraft's longitudinal axis for magnetic, true or relative bearing respectively.

bearing selector
The instrument used for selecting, or setting into an omni receiver, the VOR bearing or radial which it is desired to follow in flight.

beat frequency
The difference frequency resulting when two sinusoids are mixed in a non-linear device.

beat frequency oscillator

(1) An oscillator in a radio receiver, tuned to an intermediate frequency which beats with incoming continuous wave signals or substitutes for the carrier of single sideband signals.
(2) An early form of wide range audio oscillator which mixed the output of two RF oscillators, one fixed and one variable, to obtain an audio output.

Beaufort scale
A numerical classification of wind force.

Bellini-Tosi
Bellini and Tosi were two early experimenters who developed the direction-finding system named after them. This determined the direction of the incoming signal by comparing the relative signal strengths received on two fixed loop aerials mounted at right angles.

below minima
Weather conditions below the minimum prescribed by regulation for the particular action involved, e.g. landing minima, takeoff minima.

bench check
A functional check on aircraft components conducted in a workshop.

bending moment
The algebraic sums of all vertical forces to one side of a point.

bending stress
A combination of tension and compression stress.

Bernoulli's principle
At any point in a tube through which a fluid is flowing, static pressure and dynamic pressure equals a constant.

beryllium
A hard light metal used in making corrosion-resistant alloys. Atomic number 4, relative density 1.85, melting point 1280°C.

bias
Fabric cut diagonally across the warp and weft.

bi-directional
Refers to an interface port or bus line which can transfer data in either direction.

billion
UK 10^{12}, US 10^{9}.

bimetallic strip
Made of two metals with different coefficients of expansion, welded side by side so that changes in temperature cause bending. Used in alarm systems.

binary number system
A counting system using 2 as its base and employing the symobols 0 and 1.

biplane
An aeroplane with one wing fixed above the other.

birdstrike
Collision between aircraft and birds during the period from when the aircraft first moves under its own power with the intention of flight to when it comes to a stop at the end of the flight.

bird strike precautions
Measures taken to reduce, to the extent possible, the risk of collision between aircraft and birds during all flight phases conducted on, or in the vicinity of, aerodromes.

bisector
A straight line that divides an angle or another line into two equal parts.

bit
A unit of information which specifies one of two alternatives, i.e. 1 or 0 in binary mathematics; a unit of capacity in a store (computers).

bits per second
A measure of rate of passage of digital data.

bitumen
Many mixtures of viscous hydrocarbons are called bitumens, e.g. tar, asphalt, etc.

black blizzard
A dust storm.

black box
Flight recorder consisting of flight data recorder and a cockpit voice recorder.

blade angle
The varying angle of incidence of a propeller blade from its root to its tip.

blade profile
The form or outline of a rotor blade section.

blade radius
The radius of a circle described by the tip of a rotating blade, i.e. the distance from the blade tip to the axis of rotation.

blade stall (helicopter)
A condition which can occur when a rotor blade operates at an angle of attack greater than the angle of maximum lift.

blade twist (helicopter)
The variation or twist in the blade angle built into a rotor blade, usually such that the angle decreases towards the blade tip.

blade width ratio (helicopter)
The ratio of the width of a rotor blade at a given section to the diameter of the circle described by the tip of the rotating blade.

blast fence
A physical barrier that is used to divert or dissipate jet or propeller blast.

bleed air
Compressed air from the engine compressor directed into the cabin pressurisation system, or to drive other services.

blind speed
The rate of departure or closing of a target relative to the radar antenna at which cancellation of the primary radar target by moving target indicator (MTI) circuits in the radar equipment causes a reduction or complete loss of signal.

blind spot
An area from which radio transmissions and/or radar echoes cannot be received. The term is also used to describe portions of an airport not visible from the control tower.

blind transmission
A transmission from one station to another in circumstances where two-way communication cannot be established but where it is believed that the called station is able to receive the transmission.

blind velocity [ICAO]
The radial velocity of a moving target such that the target is not seen on primary radars fitted with certain forms of fixed echo suppression.

blind zone   See BLIND SPOT.

blizzard
A severe weather condition characterised by low temperatures and strong winds bearing a great amount of snow, either falling or picked up from the ground.

block to block time
The period between the time the aircraft first commences to move with the intention of flight to the time it finally comes to a stop after the flight.

blowing dust
A type of lithometeor composed of dust particles picked up locally from the surface and blown about in clouds or sheets.

blowing sand
A type of lithometeor composed of sand picked up locally from the surface and blown about in clouds or sheets.

blowing snow
A type of hydrometeor composed of snow picked up from the surface by the wind and carried to a height of 6 ft or more.

blowing spray
A type of hydrometeor composed of water particles picked up by the wind from the surface of a large body of water.

blushing
Whitish bloom on doped aircraft surface caused by draught or damp atmosphere, or too rapid solvent evaporation when dope is applied.

Bohr theory
A theory about the distribution of sub-atomic particles. It has now been supplanted by wave mechanics.

boiling point
The temperature at which the maximum vapour pressure of a liquid equals or exceeds the vapour pressure of the surrounding gas.

Boltzmann's constant
$R/L = 1.380622 \times 10^{-23}$ J/K° where R = the gas constant and L = Avogadro's constant.

bond
Bonded store in which duty free goods are kept under strict Customs security.

bonded fuel
Aviation fuel imported into a country for use only in international operations, on which no taxes are levied.

bonding
Connecting all metal parts of the aircraft to secure good electrical continuity and so avoid the undesirable effects of 'static' electricity.

Boolean algebra
Logical processes in computers using operators such as 'and', 'not-and', 'or', etc.

booster
First stage rocket.

bora
Katabatic wind off the North Mediterranean Sea

bottom dead centre
That point where a piston reaches the lowest limit during its stroke.

boundary layer
Layer of fluid closest to the surface over which it is flowing having a reduced rate of flow because of adhesion between the fluid and the surface.

boundary layer control
Method used to control boundary layer flow with the object of reducing drag to a minimum and thus improve the performance of an aircraft.

boundary layer thickness
The measurement of thickness of the boundary layer. The thickness varies dependent upon its distance from the leading edge of the wing.

boundary lights   See AIRPORT LIGHTING.

Boyle's law
At constant temperature, the volume of a given weight of gas is inversely proportional to its pressure, PV = constant.

bracket (landmark)
A selected pair of prominent landmarks, one on either side of the route being flown, which would make unintentional departure from the route virtually impossible.

brake horsepower
The power delivered at the propeller shaft (main drive or main output) of an aircraft engine.

braking action
A report of conditions on the airport movement area providing a pilot with a degree/quality of braking that he might expect. Braking action is reported in terms of good, medium, fair, poor or nil. See also RUNWAY CONDITION READING.

braking action advisories
When tower controllers have received runway braking action reports which include the terms 'poor' or 'nil', or whenever weather conditions are conducive to deteriorating or rapidly changing runway braking conditions, the tower should include on the ATIS broadcast the statement 'Braking action advisories are in effect'. During the time braking action advisories are in effect, ATC will issue the latest braking action report for the runway in use to each arriving and departing aircraft. Pilots should be prepared for deteriorating braking conditions and should request current runway condition information if not volunteered by controllers. Pilots should also be prepared to provide a descriptive runway condition report to controllers after landing.

break
Used to separate a series of nonstop clearances, e.g., 'Cessna 95K, turn left heading three six zero, break, TWA 105, cleared for visual approach 19L, break'.

break away height
The minimum height below which an approach cannot be continued.

bright band
In radar meteorology, a narrow, intense echo on the range – height indicator scope, resulting from water-covered ice particles of high reflectivity at the melting level.

bright display
A radar display capable of being used under relatively high ambient light levels.

British thermal unit
Heat gained or lost when the temperature of 1 lb of water changes by $1°F = 251.997$ Cal. or 1055.06 J.

broadcast
Transmission of information for which an acknowlegement is not expected.

broadcast [ICAO]
A transmission of information relating to air navigation that is not addressed to a specific station or stations.

bug
A fault, usually in software; a mark, fixed or set, on a meter (instrument) face.

bulkhead
A transverse partition which completely separates one compartment from another. The most common example is the fireproof bulkhead behind the engine.

burn off
Amount of fuel consumed by an aircraft.

bus
One or more conductors used as an information path. A conductor used to carry a particular power supply to various user equipments.

Buys Ballot's law
If an observer in the northern hemisphere stands with his back to the wind, lower pressure is to his left.

by-pass ratio
The ratio of the air mass flow through the by-pass ducts of a gas turbine engine to the air mass flow through the combustion chambers calculated at maximum thrust when the engine is stationary in an International Standard Atmosphere at sea level.

byte
A specific number of bits (usually 8) treated as a group: 8-bit word.

C

cabin altitude
The pressure altitude which corresponds to the pressure inside an aircraft cabin.

cabin water spray system
An on board system which produces a water mist throughout the aircraft cabin. The water mist can fire harden the fuselage structure to an extent that penetration of an external fire through the skin of an aircraft fuselage into the cabin can be delayed. Also, the water mist can limit the fire propagation within the cabin by the absorption of radiant and convective heat, thus preventing the occurrence of a flash fire. Additionally the water drops are sized to remove water soluble gases and wash out solid particulates from a fire to maintain a survivable atmosphere and extend the evacuation time.

cabotage [ICAO]
With respect to the Convention on International Civil Aviation each contracting State shall have the right to refuse permission to the aircraft of other contracting States to take on in its territory passengers, mail and cargo carried for remuneration or hire and destined for another point within its territory. Each contracting State undertakes not to enter into any arrangements which specifically grant any such privilege on an exclusive basis to any other State or an airline of any other State, and not to obtain any such exclusive privilege from any other State.

calculus
A branch of mathematics subdivided into: differential calculus which deals with continuously varying quantities; and integral calculus which deals with the summation of infinitely small elements.

calibrate
To adjust the scale (graduations) of a measuring instrument or device.

calibrated airspeed [US]
Indicated airspeed of an aircraft, corrected for position and instrument error. Calibrated airspeed is equal to true airspeed in standard atmosphere at sea level. In the United Kingdom the term used is 'rectified airspeed'.

call sign
Letters, numbers or words used for identification purposes.

call-up
Initial voice contact between a facility and an aircraft, using the identification of the unit being called and the unit initiating the call.

calm
The absence of apparent motion of the air.

calorie
A unit of heat which equals the heat lost or gained when the temperature of 1 gramme of water changes by 1°C = 4.1855J.
— the international table calorie = 4.1868J.
— the K cal = 1000 calories.

calorific value
Heat produced by the complete combustion of a given mass of fuel.

camber
Curvature of the surface of an aerofoil.

camfax
Facsimile transmission network for meteorological charts.

canard
Aircraft designed with the normal horizontal tail surface on the front instead of the rear of the fuselage.

candela (cd)
SI unit of luminous intensity.

canopy
Transparent cover designed to slide backwards or forwards, or hinge upwards to allow aircraft occupants to embark or disembark.

cantilever
A gear or beam fixed rigidly at one end only.

capacitance
The property of a system of conductors and insulators (a system known as a capacitor) which allows the storage of an electric charge when a p.d. exists between the conductors. In a capacitor, the conductors are known as electrodes or plates, and the insulator, which may be solid, liquid or gaseous, known as the dielectric.

capacitor
A system of conductors and insulators which store electrical charge. The SI unit is the farad.

capacitor discharge light
A lamp in which high-intensity flashes of extremely short duration are produced by the discharge of electricity at high voltage through a gas enclosed in a tube.

capacity
Ability, expressed in numbers of aircraft entering a specified portion of the airspace in a given period of time, of the ATC system or any of its sub-systems or an operating position to provide service to aircraft during normal activities. The maximum peak capacity which may be achieved for short periods may be appreciably higher than the sustainable value.

cap cloud
A standing or stationary cap-like cloud crowning a mountain summit. Also called CLOUD CAP.

capsule
An evacuated airtight container used to detect changes in pressure.

captive balloon
A balloon which, when in flight, is attached by a retaining device to the surface.

carbon (C)
Atomic number 6, atomic weight 12.011, melting point 3550°C. Conducts electricity. (Allotropes: diamond relative density 3.51; graphite relative density 2.25).

carbon dioxide (CO_2)
A colourless gas resulting from the burning of substances containing carbon. Exhaled by humans and animals; absorbed by plants; used in fire-extinguishers and 'fizzy' drinks. It freezes into a solid called 'dry ice' at −78.5°C.

carbon fibre
Threads of absolutely pure carbon in which the crystallites are orientated along the axis of the fibre. At 600 000 fibres per cm^2 cross-section, they are used to reinforce ceramics, metals or even resin. These composite materials are now used in jet and rocket engines where great strength is required at high temperatures.

carbon monoxide (CO)
The result of incomplete combustion of fuels. It is very dangerous because it is both inflammable and poisonous.

carburettor
A device which uses a venturi to vaporise petrol and mix it with the air passing through the venturi to the cylinders of an internal combustion engine.

cardinal altitudes
Odd or even thousand foot altitudes or flight levels, e.g. 5000, 6000, 7000, FL 250, FL 260, FL 270. See also ALTITUDE, FLIGHT LEVELS.

cardinal flight levels   See CARDINAL ALTITUDES.

cargo
Any property carried on an aircraft other than mail, stores and baggage.

cargo aircraft
Any aircraft, other than a passenger aircraft, which is carrying goods or property.

carrier wave
The continuous transmission of a wave of constant amplitude and frequency. This is modulated by oscillating electric current impulses.

cartesian co-ordinates
Fixing position by reference to an ordinate (horizontal x − axis) and an abscissa (vertical y − axis) and, sometimes, a third dimension, or z − axis.

category
(1) As used with respect to the certification, ratings, privileges and limitations of airmen, a broad classification of aircraft. Examples include: aeroplane, rotorcraft, glider and lighter-than-air.

(2) As used with respect to the certification of aircraft, a grouping of aircraft based upon intended use or operating limitations. Examples include: transport, normal, utility, aerobatic, limited and restricted.

category A (US)
With respect to transport category rotorcraft, multi-engine rotorcraft designed with engine and system isolation features specified in Federal Administration Regulations and utilising scheduled takeoff and landing operations under a critical engine failure concept which assures adequate designated surface area and adequate performance capability for continued safe flight in the event of engine failure.

category B (US)
With respect to transport category rotorcraft, single-engine or multi-engine rotorcraft which do not fully meet all Category A standards. Category B rotorcraft have no guaranteed stay-up ability in the event of engine failure, and unscheduled landing is assumed.

category I operations
With respect to the operation of aircraft, a straight-in approach to the runway of an airport under a Category I ILS instrument approach procedure issued by the administrator or other appropriate authority.

category II operations
With respect to the operation of aircraft, a straight-in ILS approach to the runway of an airport under a Category II ILS instrument approach procedure authorised by the appropriate civil aviation authority.

category III operations
With respect to the operation of aircraft, an ILS approach to, and landing on, the runway of an airport using a Category III ILS instrument approach procedure authorised by the appropriate civil aviation authority.

cathode
Negative pole.

cathode-ray direction
An automatic direction finder, usually VHF or UHF, which incorporates a cathode-ray tube for bearing display.

cathode-ray oscilloscope
Produces a visible image of varying electrical quantities and used as an indicator in a radar system.

cathode rays
Streams of electrons emitted from the cathode in a (near) vacuum tube.

cathode-ray tubes
Television screens, oscilloscopes, comprising a (near) vacuum tube in which a stream of electrons from the cathode pass between (and are deflected by) vertical and horizontal deflection plates before causing a pattern of luminescence on the screen.

cation
Postively charged ion which is attracted to the cathode.

cavitation
The formation of cavities in fluids caused by low pressure/high velocity (Bernoulli's Principle). On reaching regions of high pressure their collapse impact can cause damage to propellers.

ceiling [US]
The height above the earth's surface of the lowest layer of clouds or obscuring phenomena that is reported as 'broken', 'overcast' or 'obscuration', and not classified as 'thin' or 'partial'.

ceiling [ICAO]
The height above the ground or water of the base of the lowest layer of cloud below 6000 m (20 000 ft) covering more than half the sky.

ceiling and visibility OK
A term used in meteorological forecasts and reports for aviation: visibility 10 km or more; no cloud below 5000 ft, or below the minimum sector attitude, whichever is the greater, and no cumulonimbus; no precipitation reaching the ground, thunderstorm, shallow fog or low drifting snow.

ceiling and visibility unlimited [US]
A term used in meteorological forecasts and reports for aviation: clear, or scattered clouds, and visibility greater than 10 miles.

ceiling balloon
A small balloon used to determine the height of a cloud base or the extent of vertical visibility.

ceiling light
An instrument which projects a vertical light beam onto the base of a cloud or into surface-based obscuring phenomena. Used at night in conjunction with a clinometer to determine the height of the cloud base or as an aid in estimating the vertical visibility.

ceilometer
A cloud-height measuring system. It projects light onto the cloud, detects the reflection by a photo-electric cell and determines height by triangulation.

celestial navigation
The determination of geographical position by reference to celestial bodies. Normally used in aviation as a secondary means of position determination.

cell
(1) A chemical device for producing electric current. In a primary cell the chemical reaction is irreversible and the cell cannot be recharged. In a secondary cell the chemical reaction is designed to be reversible so that the battery may be recharged.
(2) A circuit or device used for storing one character or word, the location being given by a particular address.

Celsius temperature scale (centigrade)
Uses 0° for the melting point of ice and 100° for the boiling point of water. These degrees have been used for the kelvin scale except that absolute zero −273.15°C = 0°K and 273.15°K = 0°C.

center's area [US]
The specified airspace within which an air route traffic control centre (ARTCC) provides air traffic control and advisory service. See also AIR ROUTE TRAFFIC CONTROL CENTRE.

centigrade temperature scale   See CELSIUS TEMPERATURE SCALE.

central control function [UK]
An ATC system used in the LTMA to increase the capacity of air traffic flow by the use of discrete airspace sectors (tunnels in the sky concept).

central processing unit
A device capable of executing instructions obtained from memory or other sources; a term often used for a microprocessor.

centreline
A real or imaginary line down the middle of the runway.

centre of gravity
The centre of mass from which the sum of all moments of inertia is zero. The balancing point.

centre of lift
The resultant of the centre of pressure acting upon a wing.

centre of pressure
That point on the chord line at which the total lift of a section of an aerofoil is considered to act. The sum of all the moments of the pressure forces about the centre of pressure is zero.

centrifugal force
An imaginary force invented for convenience to balance the real centripetal force.

centrifugal twisting moment
The moment during a propeller's rotation which tends to twist the blades to a lower angle.

centripetal force
Mass × the inwards acceleration of bodies following curved paths = mv^2/r.

cermet
A mixture of ceramic and metal with a very high resistance to heat, corrosion and abrasion.

certificate of airworthiness
An aircraft certificate issued by an aviation authority before an aircraft is permitted to be used in service.

certificate of clearance [UK]
A certificate that forms part of the requirement for aircraft flown to specified operating conditions.

certificate of maintenance review [UK]
A certificate issued by the airworthiness authority following a review of the current maintenance records of aircraft with a certificate of airworthiness in either the transport or aerial work category.

certificate of release to service [UK]
A document certifying that those tasks which have been completed have been carried out in an approved manner.

chaff
Strips of coated material, usually aluminium foil, sometimes used for alerting radar controllers to the presence of aircraft in distress and with two way radio failure. Packets of chaff when dropped in flight are scattered by the slipstream and provide a distinctive radar pattern which should be sufficient to attract the attention of controllers.

chandelle
A flight manoeuvre in which air speed is traded to gain altitude while turning through 180°.

change of state
In meteorology, the transformation of water from one form, i.e. solid (ice), liquid or gaseous (water vapour) to any other form. There are six possible transformations designated by the following five terms:

(1) Condensation. The change of water vapour to liquid water.
(2) Evaporation. The change of liquid water to water vapour.
(3) Freezing: The change of liquid water to ice.
(4) Melting. The change of ice to liquid water.
(5) Sublimation. The change of: (a) ice to water vapour, or (b) water vapour to ice.
    See also LATENT HEAT.

changeover point
The point along an airway, usually approximately halfway between two VORs, at which it is expected that a pilot will change from the station behind to the station ahead.

channel
For the transmission signals on a designated frequency band. Channel capacity is the maximum number of signals that can be transmitted per second.

character
A unit of information in computers comprising six bits.

Charles's law
At constant pressure the volume of a given mass of gas is directly proportional to its absolute temperature (K) i.e., the gas expands by $1/273 \times$ volume at 0°C for each degree C rise in temperature.

charted visual flight procedure (US) approach
An approach wherein a radar-controlled aircraft on an IFR flight plan, operating in VFR conditions and having an ATC authorisation may proceed to the airport of intended landing via visual landmarks and altitudes depicted on a visual flight procedure chart.

charts
Navigational maps prepared by recognised organisations.

chase
An aircraft flown in proximity to another aircraft normally to observe its performance during training or testing.

chase aircraft   See CHASE.

check list
A printed list indicating actions by flight crew members in precise order of operation.

check point
A prominent landmark on the ground used to establish the position of an aircraft in flight.

Chinook
A foehn wind blowing down the eastern slope of the Rocky Mountains over the adjacent plains in the United States and Canada. In winter this warm, dry wind causes snow to disappear with remarkable rapidity, and hence it has been nick-named the 'snow-eater'.

chip
A collection of interconnected electronic components formed on a single silicon wafer.

choking coil
A high inductance but low resistance coil used to suppress a.c. when d.c. is required in the circuit.

chord
(1) A straight line joining two points on a curve.
(2) A straight line from the leading edge to the trailing edge of an aircraft wing.

chromium (Cr)
Atomic number 24, relative density 7.18, melting point 1857°C. Bright metal used for plating and stainless steel.

chronometer
An exceptionally precise clock, watch or other timepiece.

circle, great   See GREAT CIRCLE.

circle, small   See SMALL CIRCLE.

circle-to-land
A manoeuvre initiated by the pilot to align the aircraft with a runway for landing when a straight-in landing from an instrument approach is not possible or is not

desirable. The manoeuvre is made only after ATC authorisation has been obtained and the pilot has established required visual reference to the airport. See also CIRCLE TO RUNWAY and LANDING MINIMUMS.

circle to runway (runway numbered)
Used by ATC to inform the pilot that he must circle to land because the runway in use is other than the runway aligned with the instrument approach procedure. When the direction of the circling manoeuvre in relation to the airport/runway is required, the controller will state the direction (eight cardinal compass points) and specify a left or right downwind or base leg as appropriate, e.g. 'Cleared VOR runway three six approach circle to runway two two' or 'circle northwest of the airport for a right downwind to runway two two'. See also CIRCLE-TO-LAND and LANDING MINIMUMS.

circling approach
An extension of an instrument approach procedure which provides for visual circling of the aerodrome prior to landing.

circuit breaker
A device which takes the place of a fixed fuse in an electric circuit. In event of an overload, the circuit breaker 'pops out' and can be re-set by pushing it in.

circulation
The motion of air along a curved path.

cirriform
All species and varieties of cirrus, cirrocumulus and cirrostratus clouds; describes clouds composed mostly or entirely of small ice crystals, usually transparent and white. Cirriform clouds often producing halo phenomena not observed with other cloud forms. Average height ranges upward from 20 000 ft in middle latitudes.

cirrocumulus
A cirriform cloud appearing as a thin sheet of small white puffs resembling flakes or patches of cotton without shadows; sometimes confused with altocumulus.

cirrostratus
A cirriform cloud appearing as a whitish veil, usually fibrous, sometimes smooth and which often produces halo phenomena. May totally cover the sky.

cirrus
A cirriform cloud in the form of thin, white feather-like clouds in patches or narrow bands. The clouds have a fibrous and/or silky sheen, large ice crystals often trailing downward a considerable vertical distance in fibrous slanted or irregularly curved wisps called mares' tails.

civil aircraft [ICAO]
Aircraft other than public aircraft. See also PUBLIC AIRCRAFT.

(1) Category of aircraft
   As used with respect to the certification of aircraft, a broad grouping of aircraft having similar characteristics i.e. aeroplane, helicopter, glider, free balloon.
(2) Class of aircraft
   As used with respect to the certification, ratings, privileges and limitation of

flightcrew, refers to a classification of aircraft within a category having similar operating characteristics, i.e. single-engine, multi-engine, land, water.

civil tilt rotor
An aircraft with the characteristics of a helicopter and a fixed-wing aircraft.

clarifier
Alternative nomenclature for BFO tuning control on some receivers designed primarily for SSB reception.

clear air turbulence
Turbulence encountered in air where no clouds are present, more popularly applied to high level turbulence associated with wind shear. Turbulence of any sort is more generally classified as follows:
(1) Light. Causes slight, rapid and somewhat rhythmic bumpiness without appreciable changes in altitude or attitude. No IAS fluctuations. Reported as light chop.
(2) Moderate. Turbulence which is similar to light turbulence but of greater intensity. Changes in altitude or attitude occur but the aircraft remains in positive control at all times. IAS fluctuates 15 – 25 kts. Reported as moderate turbulence.
(3) Severe. Turbulence that causes large, abrupt changes in altitude and/or attitude. Aircraft may be temporarily out of control. IAS fluctuates more than 25 kts. Reported as severe turbulence.

clearance function
The formulation and transmission of a clearance by an air traffic control unit as well as the acknowledgement and acceptance of such clearance by the pilot.

clearance limit
The fix, limit or location to which an aircraft is cleared when issued with an air traffic clearance.

clearance limit [ICAO]
The point at which an aircraft is granted an air traffic control clearance.

clearance void (time)
Used by ATC to advise an aircraft that the departure clearance is automatically cancelled if takeoff is not made prior to a specified time. The pilot must obtain a new clearance or cancel his IFR flight plan if not off by the specified time.

clearance void time [ICAO]
A time specified by an air traffic control unit at which a clearance ceases to be valid unless the aircraft concerned has already taken action to comply therewith.

cleared as filed
The aircraft is cleared to proceed in accordance with the route of flight filed in the flight plan. This clearance does not include the altitude, SID or SID Transition. See also REQUEST FULL ROUTE CLEARANCE.

cleared for approach
ATC authorisation for an aircraft to execute any standard or special instrument approach procedure for that airport. Normally an aircraft will be cleared for a specific instrument approach procedure. See also CLEARED FOR (TYPE OF) APPROACH.

cleared for takeoff
ATC authorisation for an aircraft to depart. It is predicated on known traffic and known physical airport conditions.

cleared for the option [US]
ATC authorisation for an aircraft to make a touch-and-go, low approach, missed approach, stop and go or full stop landing at the discretion of the pilot. It is normally used in training so that an instructor can evaluate a student's performance under changing situations. See also OPTION APPROACH.

cleared for (type of) approach
ATC authorisation for an aircraft to execute a specific instrument approach procedure to an airport, e.g. 'Cleared for ILS runway three six approach'.

cleared through
ATC authorisation for an aircraft to make intermediate stops at specified airports without refiling a flight plan while en-route to the clearance limit.

cleared to land
ATC authorisation for an aircraft to land. It is predicated on known traffic and known physical conditions of the airport.

clear ice   See CLEAR ICING.

clear icing
Generally, the formation of a layer or mass of ice which is relatively transparent because of its homogeneous structure, and small number and size of air spaces. Used commonly as synonymous with glaze, particularly with respect to aircraft icing. Compare with rime icing. Factors which favour clear icing are large drop size, such as those found in cumuliform clouds, rapid accretion of super-cooled water, and slow dissipation of latent heat of fusion.

clearway [ICAO]
A defined rectangular area on the ground or water under the control of the appropriate authority, selected or prepared as a suitable area over which an aeroplane may make a portion of its initial climb to a specified height.

clearway [UK]
An area at the end of the takeoff run available and under the control of the aerodrome licensee, selected or prepared as a suitable area over which an aircraft may make a portion of its initial climb to a specified height.

climate
The prevalent or characteristic meteorological conditions of any place or region, and their extremes.

climatology
The study of climate.

climb
That period of flight during which an aircraft is climbing to its intended cruising altitude. The rate of climb is generally expressed in feet per minute, measured by a vertical speed indicator.

climb gradient
Height gained expressed as a percentage of horizontal distance travelled.

climbout
That portion of flight operation between takeoff and the initial cruising altitude.

climb phase
The operating phase defined by the time during which the engine is operated in the climb operating mode.

climb to VFR [ICAO]
Authorisation for an aircraft to climb to Visual Flight Rules (VFR) conditions within a control zone when the only weather limitation is restricted visibility. The aircraft must remain clear of clouds while climbing to VFR. See also SPECIAL VFR.

clinometer
An instrument used in weather observing for measuring angles of inclination; it is used in conjunction with a ceiling light to determine cloud height at night.

closed flight plan
The closure of a flight plan when the aircraft's safe arrival has been reported to air traffic control.

closed runway
A runway that is unusable for aircraft operations. Only the airport management/ military operations office can close a runway.

closed traffic [US]
Successive operations involving takeoffs and landings or low approaches where the aircraft does not exit the traffic pattern.

closure rate
The relative closing velocity between two aircraft.

cloud bank
Generally, a fairly well-defined mass of cloud observed at a distance; it covers an appreciable portion of the horizon sky, but does not extend overhead.

cloud banner   See BANNER CLOUD.

cloud base recorder   See CEILOMETER.

cloud burst
In popular terminology, any sudden and heavy fall of rain, almost always of the shower type.

cloud cap   See CAP CLOUD.

cloud ceiling [UK]
In relation to an aerodrome, the distance measured vertically from the notified elevation of the aerodrome to the lowest part of any cloud visible from the aerodrome which is sufficient to obscure more than one half of the sky so visible.

cloud detection radar
A vertically directed radar to detect cloud bases and tops.

clutch (helicopter)

The clutch of a typical helicopter's rotor system allows a stress-free operation. At slow rotor speeds when there is insufficient centrifugal force to hold down the rotor blades in a normal flat disc position, strong or gusting winds may be able to raise the rotor blades well above their normal position and than suddenly drop them, possibly causing much damage to them and endangering the aircraft. All helicopters, therefore, have been designed with some form of anti-coning device to help prevent this possibility.

clutter

In radar operations clutter refers to the reception and visual display of radar returns caused by precipitation, chaff, terrain, numerous aircraft targets or other phenomena. Such returns may limit or preclude ATC from providing services based on radar. See also CHAFF, GROUND CLUTTER, PRECIPITATION and TARGET.

coanda effect

A tendency for air to follow the curvature of a surface.

coastal fix

A navigation aid or intersection where an aircraft transits between the domestic route structure and the oceanic route structure.

co-axial cable

A central conducting wire surrounded by a dielectric (polythene), inside a cylindrical outer conductor. Used to transmit high frequency signals.

cockpit voice recorder

A tape recording system fitted to an aircraft which records voice communication between the flight crew. When required to be carried by aviation regulations, flight recorders must be constructed, located and installed so as to provide maximum practical protection for the recordings in order that the recorded information may be preserved, recovered and transcribed.

code

(1) The series of pulses by which an airborne transponder replies to interrogation from the ground.

(2) A system of symbols and rules used for representing information such as numbers, letters and control signals.

cold front

Any non-occluded front which moves in such a way that colder air replaces warmer air.

cold wave [US]

A rapid and marked fall of temperature. The US Weather Bureau applies this term to a fall of temperature in 24 hours equalling or exceeding a specified number of degrees and reaching a specified minimum temperature or lower. These specifications vary for different parts of the country and for different periods of the year.

collimating mark

The small cross or similar mark found in the middle of each side of a print. Used for finding a principal point.

co-location
VHF navigation and DME beacons sharing the same geographical site; such beacons will operate on paired frequencies and use associated identity.

colour perception
Defective colour perception is a permanent condition present at birth. Colour vision is tested when student pilots present themselves for an initial medical examination. On this occasion they will be confronted with a set of colour charts (Ishihara Test). If their answers are correct they will be regarded as having normal colour perception. However, should they fail, they will be subjected to a lantern test. Depending on the result of this test, they will be graded as either colour defective safe or colour defective unsafe, depending on whether or not they can distinguish accurately between red, green and white lights.

combined centre-rapcon [US]
An air traffic facility which combines the functions of an ARTCC and a radar approach control facility. See also AIR ROUTE TRAFFIC CONTROL CENTRE.

combined station/tower [US]
A control tower which also provides the services usually given by an FSS.

Comm-A
A term referring to standard-length uplink communications used in satellites.

Comm-B
A term referring to standard-length downlink communications used in satellites.

commander
In relation to an aircraft, the member of the flight crew designated as commander of that aircraft by the operator thereof, or failing such a person, the person who is for the time being the pilot in command of the aircraft.

commercial operator [US]
A person who, for compensation or hire, engages in the carriage by aircraft in air commerce of persons or property, other than as an air carrier or foreign air carrier or under the authority of FAR Part 375 of this title. Where it is doubtful that an operation is for 'compensation or hire', the test applied is whether the carriage by air is merely incidental to the person's other business or is, in itself, a major enterprise for profit.

common medium term plan [EUROCONTROL]
A planning document developed by EUROCONTROL for the purpose of achieving an efficient air traffic service within the participating States of EUROCONTROL.

common portion [US]   See COMMON ROUTE (US).

common route [US]
That segment of a North American route between the inland navigation facility and the coastal fix.

communication centre
An aeronautical fixed station which relays or re-transmits telecommunications traffic from, or to, a number of other aeronautical fixed stations directly connected to it.

communications satellite
A man-made communications station that orbits the earth. A passive satellite reflects radio and television signals, etc.; an active satellite receives and transmits signals.

commutator
An arrangement of insulated metal bars connected to the coils of an electric motor or generator to produce an unidirectional current from the generator or a reversal of current into the coils of the motor.

compass calibration pad
A suitable platform at principal airports, designed for checking the directions indicated by the compass of an aircraft against known magnetic directions, and so determining the error or deviation of the compass.

compass compensation
The systematic reduction of compass deviation by adjusting small magnets incorporated in a magnetic compass for that purpose.

compass locator
A low power, low or medium frequency (L/MF) radio beacon installed in conjunction with the outer or middle marker of an instrument landing system (ILS). It can be used for navigation at distances of approximately 15 miles or as authorised in the approach procedure. See also MIDDLE COMPASS LOCATOR (LMM) and OUTER COMPASS LOCATOR (LOM).

compass, magnetic   See MAGNETIC COMPASS.

compass rose
A circle, graduated in degrees from 0 to 360, printed on aeronautical charts as a reference to directions, true or magnetic.

Competent ATS authority
In relation to the United Kingdom, the Civil Aviation Authority, and in relation to any other country the authority responsible under the law of that country for promoting the safety of civil aviation.

complementary angles
Angles totalling 90°.

component forces or velocities
When a combination of two or more forces or velocities produces a single effect – the 'resultant'.

composite flight plan
A flight plan that specifies VFR operation for one portion of flight and IFR for another portion.

composite route system
An organised oceanic route structure incorporating reduced lateral spacing between routes, in which composite separation is authorised.

composites
A term used to describe compounds that, in combination, have properties superior to those of conventional materials such as steel or aluminium. This can include, but is certainly not limited to, plastics, fibreglass and other glass products, graphite, Kevlar, titanium and ceramics. The term 'advanced composites' refers to compounds such as graphite and Kevlar, which have enough stiffness to allow them to be used in manufacturing airframe fuselages and other primary structures. Advanced composites are more structurally efficient and able to handle heavily loaded applications.

composite separation
A method of separating aircraft in a composite route system where, by management of route and altitude assignments, a combination of half the lateral minimum specified for the area concerned and half the vertical minimum is applied.

compressibility
The natural ability of air to change volume when compressed or expanded.

compressibility effects
(1) The formation of shock waves when the aircraft moves at very high speeds.
(2) In relation to an airspeed indicator, the error of the indicated air speed due to the compression of air at the pitot-static tube of high speed aircraft.

compulsory reporting points
Reporting points that must be reported to ATC. They are designated on aeronautical charts by solid triangles or filed in a flight plan as fixes selected to define direct routes. These points are geographical locations that are defined by navigation aids/fixes. Pilots should discontinue position reporting over compulsory reporting points when informed by ATC that their aircraft is in 'radar contact'.

computed centre line approach
An MLS approach in which the precision segment is based on a computed path along the extended runway centreline which is not aligned with an MLS radial and/or MLS elevation angle.

computer
A device which performs sequences of arithmetical and logical steps upon data without human intervention.

computer aeronautical
A device for graphic solution of wind triangles and other problems of dead reckoning, including speed – time – distance problems. See also COURSE LINE COMPUTER.

condensation
The change of a vapour to a liquid when the vapour pressure of the gas falls to that of the liquid at that temperature.

condensation level
The height at which a rising column of air reaches saturation, and clouds form.

condensation nuclei
Small water-absorbent particles in the air on which water vapour condenses or sublimates.

condensation trail
A cloud-like streamer (also called a contrail or vapour trail) frequently observed to form behind aircraft flying in clear, cold, humid air.

condenser
Electrical capacitor.

conditional instability   See CONDITIONAL STABILITY.

conditional stability
The state of a column of air when its vertical distribution of temperature is such that the layer is stable for dry air, but unstable for saturated air.

conditionally unstable air
Unsaturated air that will become unstable when it becomes saturated. See also INSTABILITY.

conductance (G)
The reciprocal of resistance. The unit is the siemens.

conduction
The transfer of heat by molecular action through a substance or from one substance in contact with another. Transfer is always from warmer to colder temperature.

conference communications
Communications facilities whereby direct-speech conversation may be conducted between three or more locations simultaneously.

configuration (as applied to an aeroplane)
A particular combination of the positions of the moveable elements, such as wing flaps, landing gear, etc., which affect the aerodynamic characteristics of the aeroplane.

conflict
Predicted converging of aircraft in space and time which constitutes a violation of a given set of separation minima.

conflict alert
A function of certain air traffic control automated systems designed to alert radar controllers to existing or pending situations recognised by the programme parameters that require their immediate attention/action.

conflict detection
The discovery of a conflict as a result of a conflict search.

conflict resolution
The determination of alternative flight paths which would be free from conflict and the selection of one of these flight paths for use.

conflict search
Computation and comparison of the predicted flight paths of two or more aircraft for the purpose of determining conflicts.

conformality
The property of a chart to present angles, bearings, etc., correctly. This is an essential requirement for navigation charts.

consol
Low/medium frequency navaid.

consolan
A system of radiobeacon signals developed from the German 'Sonne'. Simply counting dots and dashes heard with an ordinary receiver, bearings may be determined quite accurately at distances up to 1000 miles, or more. Essentially the same as the British 'Consol'. Used chiefly in transoceanic navigation.

constant pressure chart
A chart of a constant pressure surface; may contain analyses of height, wind, temperature, humidity and/or other elements.

contact
(1) Establish communication with (followed by the name of the facility and, if appropriate, the frequency to be used).
(2) A flight condition wherein the pilot ascertains the attitude of his aircraft and navigates by visual reference to the surface. See also CONTACT APPROACH and RADAR CONTACT.

contact approach [US]
An approach wherein an aircraft on an IFR flight plan, having an air traffic control authorisation, operating clear of clouds with at least one mile flight visibility and a reasonable expectation of continuing to the destination airport in those conditions, may deviate from the instrument approach procedure and proceed to the destination airport by visual reference to the surface. This approach will only be authorised when requested by the pilot and the reported ground visibility at the destination airport is at least one statute mile.

contact point
A specified position, time or level at which an aircraft is required to establish radio communication with an air traffic control unit.

continental polar air   See POLAR AIR.

continental tropical air   See TROPICAL AIR.

continuous wave
Radio or radar transmission transmitted continuously and not in separate pulses.

contour
An imaginary line formed by the intersection of a horizontal plane with the surface of the earth, all points on any given contour being at the same elevation with respect to sea level (or other chosen reference plane); a line of equal height on a constant

pressure chart, analogous to contours on a relief map; in radar meteorology, a line on a radar scope of equal echo intensity.

contouring circuit
On weather radar, a circuit which displays multiple contours of echo intensity simultaneously on the plan position indicator or range – height indicator scope. See also CONTOUR.

contour interval
The vertical separation between the horizontal planes of two adjacent contours.

contrail
A cloud-like streamer frequently observed behind aircraft flying in clear, cold, humid air caused by the addition to the atmosphere of water vapour from engine exhaust gases.

control area
A controlled airspace extending upwards between specified limits above the earth.

control assistant
A person who assists in the provision of air traffic services but who is not authorised to make decisions regarding clearances, advice or information to be issued to aircraft.

controlled aerodrome [ICAO]
An aerodrome at which air traffic control service is provided to aerodrome traffic. *Note:* The term controlled aerodrome indicates that air traffic control service is provided to aerodrome traffic but does not necessarily imply that a control zone exists.

controlled airspace [ICAO]
An airspace of defined dimensions within which air traffic control services are provided to controlled flights.

controlled airspace [ICAO] **(instrument restricted)**
Controlled airspace within which only IFR flights are permitted.

controlled airspace [ICAO] **(instrument/visual)**
Controlled airspace within which IFR and controlled VFR flights are permitted.

controlled airspace [ICAO] **(visual exempted)**
Controlled airspace within which both IFR and VFR flights are permitted, but VFR flights are not subject to control.

controlled basic cover
Accurately flown basic cover for survey purposes and from which base-line photographs are produced.

controlled flight
Any flight which is provided with air traffic control service.

controlled VFR flight
A controlled flight conducted in accordance with the visual flight rules.

controller
A person authorised to provide air traffic control services.

control motion noise
That portion of a guidance signal error which causes control surface, wheel and column motion and could affect aircraft attitude angle during coupled flight, but does not cause aircraft displacement from the desired course and/or glide path.

control sector
An airspace area of defined horizontal and vertical dimensions for which a controller, or group of controllers, has air traffic control responsibility, normally within an air route traffic control centre or an approach control facility. Sectors are established based on predominant traffic flows, altitude strata and controller workload. Pilot–controller communications during operations within a sector are normally maintained on discrete frequencies assigned to the sector. See also DISCRETE FREQUENCY.

control segment
A world-wide network of GPS monitoring and control stations that ensures the accuracy of satellite positions and their clocks.

control slash [US]
A radar beacon slash representing the actual position of the associated aircraft. Normally the control slash is the one closest to the interrogating radar beacon site. When ARTCC radar is operating in narrowband (digitised) mode, the control slash is converted to a target symbol.

control system
In relation to aeroplanes, a system by which the flight path, attitude or propulsive force is changed.

control system (helicopter)
There are four principal controls common to all helicopters: cyclic pitch control (occasionally referred to as azimuth control), collective pitch control, throttle control and anti-torque tail rotor control.

control zone
A controlled airspace extending upwards from the surface of the earth to a specified upper limit.

convection
The transfer of heat from one place to another by the movements of a liquid or gas.

(1) In general, mass motions within a fluid resulting in transport and mixing of the properties of that fluid.
(2) In meteorology, atmospheric motions that are predominantly vertical, resulting in vertical transport and mixing of atmospheric properties; distinguished from advection.

convective cloud   See CUMULIFORM.

convective condensation level

The lowest level at which condensation will occur as a result of convection due to surface heating. When condensation occurs at this level, the layer between the surface and the CCL will be thoroughly mixed, temperature lapse rate will be dry adiabatic, and mixing ratio will be constant.

convective instability

The state of an unsaturated layer of air whose lapse rates of temperature and moisture are such that, when lifted adiabatically until the layer-becomes saturated, convection is spontaneous.

Convective SIGMET

A weather advisory concerning convective weather significant to the safety of all aircraft. Convective sigmets are issued for tornadoes, lines of thunderstorms, embedded thunderstorms of any intensity level, areas of thunderstorms greater than or equal to intensity level four, with an area coverage of four-tenths (40%) or more and hail three-fourths inch or greater. See also AIRMET and SIGMET INFORMATION.

convective significant meteorological information [US]   See CONVECTIVE SIGMET.

conventional current

Electrical current thought of as flowing from positive to negative (the opposite of 'electron flow').

convergence

(1) The condition that exists when the distribution of winds within a given area is such that there is a net horizontal inflow of air into the area. In convergence at lower levels, the removal of the resulting excess is accomplished by an upward movement of air; consequently, areas of low-level convergent winds are regions favourable to the occurrence of clouds and precipitation. Compare with DIVERGENCE.

(2) The angle between meridians on the surface of the earth, varying as the sine of the latitude, from 0 at the equator to 1 per degree of difference in longitude at the poles; the angle between meridians of the Lambert projection, which varies according to the standard parallels selected for the latitude band under consideration.

convergent

Coming together from different directions.

converter

A machine for converting alternating current to direct current or vice versa.

co-ordinates

Reference lines, usually vertical and horizontal, used to describe the position of a point; the intersection of lines of reference, usually expressed in degrees/minutes/seconds of latitude and longitude, used to determine position or location.

co-ordinate system (planar)

A function is said to use conical co-ordinates when the decoded guidance angle varies as the minimum angle between the surface of a cone containing the receiver antenna and a plane perpendicular to the axis of the cone and passing through its apex. The apex of the cone is at the antenna phase centre. For approach azimuth or back azimuth functions, the plane is the vertical plane containing the runway centreline. For elevation functions, the plane is horizontal.

co-ordinate system (conical)
A function is said to use planar co-ordinates when the decoded guidance angle varies as the angle between the plane containing the receiver antenna and a reference plane. For azimuth functions, the reference plane is the vertical plane containing the runway centreline, and the plane containing the receiver antenna is a vertical plane passing through the antenna phase centre.

co-ordination [ATC]
The process of obtaining agreement on clearances, transfer of control, advice or information to be issued to aircraft, by means of information exchanged between air traffic services units or between controller positions within such units.

co-ordination fix
The fix in relation to which facilities will hand off, transfer control of an aircraft or co-ordinate flight progress data. For terminal facilities it may also serve as a clearance limit for arriving aircraft.

co-pilot
A licensed pilot serving in any piloting capacity other than as a pilot-in-command, but excluding a pilot who is on board the aircraft for the sole purpose of receiving flight instruction.

copper (Cu)
Atomic number 29, melting point 1084°C, relative density 8.95. Second best conductor of electricity (after silver). Common alloys: brass and bronze.

cordite
A low explosive used in ammunition, made from cellulose nitrate and nitroglycerine.

core
Ferromagnetic rod used to increase inductance along the centre of a coil.

Coriolis force
An imaginary force used to simplify calculations concerning the effect of the earth's rotation on its atmosphere; used to replace tangential acceleration. It acts to the right of wind direction in the northern hemisphere and the left in the southern hemisphere.

Coriolis illusion (sensory)
An abrupt head movement made during a prolonged constant rate turn may set the fluid in more than one of the semicircular tubes of the vestibular apparatus in motion, creating the strong illusion of turning or accelerating in an entirely different axis. This disoriented pilot may manoeuvre the aircraft into a dangerous attitude in an attempt to correct this illusory movement.

corona
A prismatically coloured circle or arcs of a circle, with the sun or moon at its centre. Coloration is from blue inside to red outside (opposite that of a halo); varies in size

(much smaller) as opposed to the fixed diameter of the halo. A halo is characteristic of clouds composed of water droplets and valuable in differentiating between middle and cirriform clouds.

corposant   See ST ELMO'S FIRE.

corrected altitude
Indicated altitude of an aircraft altimeter corrected for the temperature of the column of air below the aircraft, the correction being based on the estimated departure of existing temperature from standard atmospheric temperature; an approximation of true altitude.

correction
An error has been made in the transmission and the correct version follows.

corrosion
Surface damage caused by chemical action of moisture, air or chemicals.

cosecant
Trigonometrical ratio: the reciprocal of the sine, i.e. hypotenuse/opposite.

cosine
Trigonometrical ratio: reciprocal of the secant, i.e. hypotenuse/adjacent.

cosmic rays
High energy radiation from space, comprising protons, electrons and alpha particles.

cotangent
Trigonometrical ratio: reciprocal of the tangent, i.e. adjacent/opposite.

coulomb (C)
Unit of charge equivalent to 6.24×10^{18} electrons.

couple
Two forces which are equal, opposite and parallel, but not colinear. Each causes a moment which is calculated as the force $\times$ the perpendicular distance to the line of action.

course
(1) The intended direction of flight in the horizontal plane measured in degrees from north. When used in aviation the word 'track' is also often used in place of 'course'.
(2) The ILS localiser signal pattern usually specified as front course or back course. See also BEARING, INSTRUMENT LANDING SYSTEMS and RADIAL.

course deviation indicator
An instrument which presents steering signals from a navaid to the pilot which, if obeyed, cause the aircraft to follow a particular flight path.

course line
The locus of points nearest to the runway centre line in any horizontal plane at which the difference in depth of modulation is zero.

course line computer
An electronic computer by means of which a VORTAC can seemingly be 'moved' to any desired location within range, where it becomes a 'phantom' station. The phantom can then be approached in the same way that a standard VORTAC might be approached, using the same flight instruments. See also AREA NAVIGATION.

course sector
A sector in a horizontal plane containing the course line limited by the loci of points nearest to the course line at which the difference in depth modulation is 0.155.

course selector   See BEARING SELECTOR.

coverage sector
A volume of airspace within which service is provided by a particular function and in which the signal power density is equal to or greater than the specified minimum.

crack
A microscopic rupture in metal which, due to repeated loads, is progressively increased in length causing the material eventually to break.

cracking (pyrolisis)
The use of heat (and sometimes a catalyst) to convert oils with a high boiling point into fuel oils with a low boiling point.

crew member
A person assigned to perform duties in an aircraft during flight time.

critical altitude
The maximum altitude at which, in standard atmosphere, it is possible to maintain, at a specified propeller RPM, a specific power or a specified manifold pressure. Unless otherwise stated, the critical altitude is the maximum altitude at which it is possible to maintain, at the maximum continuous RPM, one of the following:

(1) The maximum continuous power, in the case of engines for which this power rating is the same at sea level and at the rated altitude.
(2) The maximum continuous rated manifold pressure, in the case of engines the maximum continuous power of which is governed by a constant manifold pressure.

critical engine
In multi-engine aircraft, the engine whose failure would most adversely affect the performance or handling qualities of an aircraft.

cross bar
Approach lighting system at right angles to the line of lights forming the centreline of the runway lighting system.

cross feed
A system in multi-engine aircraft by which fuel may be transferred from one engine to another, or from one fuel tank to another.

cross (fix) at (altitude) [US]
Used by ATC when a specific altitude restriction at a specified fix is required.

cross (fix) at or above (altitude) [US]
Used by ATC when an altitude restriction at a specified fix is required. It does not prohibit the aircraft from crossing the fix at a higher altitude than specified; however, the higher altitude may not be one that will violate a succeeding altitude restriction or altitude assignment. See also ALTITUDE RESTRICTION.

cross (fix) at or below (altitude) [US]
Used by ATC when a maximum crossing altitude at a specific fix is required. It does not prohibit the aircraft from crossing the fix at a lower altitude; however, it must be at or above the minimum IFR altitude. See also ALTITUDE RESTRICTION and MINIMUM IFR ALTITUDE.

crosswind
(1) When used concerning the traffic pattern, signifies 'crosswind leg'. See also TRAFFIC PATTERN.
(2) When used concerning wind conditions, signifies a wind not parallel to the runway or the path of an aircraft. See also CROSSWIND COMPONENT.

crosswind component
The velocity component of the wind measured at or corrected to a height of 33 ft above ground level at right angles to the direction of takeoff or landing.

cruise [US]
Used in an ATC clearance to authorise a pilot to conduct flight at any altitude from the minimum IFR altitude up to and including the altitude specified in the clearance. The pilot may level off at any intermediate altitude within this block of airspace. Climb/descent within the block is to be made at the discretion of the pilot. However, once the pilot starts descent and verbally reports leaving an altitude in the block, he may not return to that altitude without additional ATC clearance. Further, it signifies approval for the pilot to proceed to and make an approach at destination airport and can be used in conjunction with:

(1) An airport clearance limit at locations with a standard/special instrument approach procedure. The FARs require that if an instrument letdown to an airport is necessary the pilot shall make the letdown in accordance with a standard/special instrument approach procedure for that airport.
(2) An airport clearance limit at locations that are within/below/outside controlled airspace and without a standard/special instrument approach procedure. Such a clearance is not authorisation for the pilot to descend under IFR conditions below the applicable minimum IFR altitude nor does it imply that ATC is exercising control over aircraft in uncontrolled airspace; however, it provides a means for the aircraft to proceed to destination airport, descend and land in accordance with applicable FARs governing VFR flight operations. Also, this provides search and rescue protection until such time as the IFR flight plan is closed.

cruise climb
(1) An aeroplane cruising technique resulting in a net increase in altitude as the aeroplane weight decreases.

(2) A climb performed by an aeroplane at a speed greater than the best rate of climb speed.

cruising altitude
An altitude or flight level maintained during en-route level flight. This is a constant altitude and should not be confused with a cruise clearance. See also ALTITUDE.

cruising level   See CRUISING ALTITUDE.

cruising level [ICAO]
A level maintained during a significant portion of a flight.

crystal microphone
A microphone in which the vibrations of sound waves cause varying pressure on a piezoelectric crystal, which in turn produces a varying emf.

culture
Generally applied to the cities, railroads, highways and other constructed features on the surface of the earth; often referred to as 'the works of man'.

cumuliform
Descriptive of clouds having dome-shaped upper surfaces in contrast to the horizontally extended stratiform types.

cumulonimbus
A cumuliform cloud type. It is heavy and dense, with considerable vertical extent in the form of massive towers. Often has tops in the shape of an anvil or massive plume. Under the base of cumulonimbus, which is often very dark, there frequently exists virga, precipitation and low ragged clouds (scud), either merged with it or not. Frequently accompanied by lightning, thunder and, sometimes, hail; occasionally produces a tornado or a waterspout. The ultimate manifestation of the growth of a cumulus cloud, occasionally extending well into the stratosphere.

cumulonimbus mamma (mammata)
A cumulonimbus cloud having hanging protuberances like pouches, festoons or udders, on the underside of the cloud. Usually indicative of severe turbulence.

cumulus
A cloud in the form of individual detached domes or towers which are usually dense and well defined, develops vertically in the form of rising mounds of which the bulging upper part often resembles a cauliflower. The sunlit part of these clouds is mostly brilliant white; their bases are relatively dark and nearly horizontal.

cumulus fractus   See FRACTUS.

current
An electric charge moving through a conductor: conventional current + to −; electron flow − to +. It is measured in coulombs per second called amperes.

current flight plan
The flight plan, including changes if any, brought about by subsequent clearances. *Note:* When the word message is used as a suffix to this term, it denotes the content and format of the current flight plan data sent from one unit to another.

cushioning effect
A temporary gain in lift during landing due to compression of air between the wings of the aircraft and the ground.

customs
A government organisation responsible for levying taxes against imported and exported items that exceed laid down duty-free allowances.

cycle
A series of changes which return a system to its original condition, e.g. alternating current. One hertz = (Hz) = one cycle per second.

cycle slip
A discontinuity in the measured carrier-beat phase resulting from a temporary loss-of-lock in the carrier loop of a GPS receiver (satellite).

cyclogenesis
Any development or strengthening of cyclonic circulation in the atmosphere.

cyclone
(1) An area of low atmospheric pressure which has a closed circulation that is cyclonic, i.e. as viewed from above, the circulation is anticlockwise in the northern hemisphere, clockwise in the southern hemisphere, undefined at the equator. Because cyclonic circulation and relatively low atmospheric pressure usually co-exist, in common practice the terms cyclone and low are used interchangeably. Also, because cyclones often are accompanied by inclement (sometimes destructive) weather, they are frequently referred to simply as storms.
(2) Frequently misused to denote a tornado.
(3) In the Indian Ocean, a tropical cyclone of hurricane or typhoon force.

cyclonite hexogen (RDX)
A high explosive.

D

daily variation
Very slight but rapid variations in the earth's magnetic field associated with sunspot activity.

Dalton's law
The total pressure of a mixture of gases is equal to the sum of the partial pressures; and the partial pressure of a gas is the pressure that it would exert if it alone filled the volume of the container.

damping factor
The ratio of peak amplitudes of successive oscillations.

damping moment
The tendency of an aircraft to return to a normal flight attitude after an upset.

danger area
An airspace of defined dimensions within which activities dangerous to the flight of aircraft may exist at specified times.

Danger area activity information service [UK]
A service which is available in respect of certain UK danger areas and which complements the present methods of promulgation of information about those danger areas which are participating in the service. The purpose of this service (DAAIS) is to enable civil pilots to obtain, via a nominated air traffic service unit (NATSU), an airborne update of the activity status of a danger area which is participating in the DAAIS and whose position is relevant to the flight of their aircraft. Such an update will assist pilots in deciding whether it would be prudent, on flight safety grounds, to penetrate the area.

dangerous goods
Articles or substances which are capable of posing a significant risk to health, safety or property when transported by air.

dangerous goods accident
An occurrence associated with, and related to, the transport of dangerous goods by air which results in fatal or serious injury to a person or major property damage.

data bus
Usually either 8 or 16 bi-directional lines capable of carrying information to and from a central processing unit, memory or interface devices of a computer system.

data convention
An agreed set of rules governing the manner or sequence in which a set of data may be combined into a meaningful communication.

data link
Signal channel along which digital messages are routed.

data message
A 1500 bit message included in the GPS signal which reports a satellite's location, clock corrections and health. Included is rough information on the other satellites in the constellation.

data processing
A systematic sequence of operations performed on data.
Note: Examples of operations are the merging, sorting, computing or any other transformation or rearrangement with the object of extracting or revising information, or of altering the representation of information.

data recorder
Electronic or electromechanical systems designed for recording data in digital or analog form for the purpose of retrieval. See also FLIGHT DATA RECORDER.

data signalling rate
Data signalling rate refers to the passage of information per unit of time, and is expressed in bits/second. Data signalling rate is given by the formula:

$$\sum_{i=1}^{i=m} \frac{1}{T_i} \log 2^{n_i}$$

where m is the number of parallel channels, T_i is the minimum interval for the ith channel expressed in seconds, n_i is the number of significant conditions of the modulation in the ith channel.
Note 1: For a single channel (serial transmission) it reduces to $(1/T) \log_2 n$; with a two-condition modulation $(n = 2)$, it is $1/T$.
  For a parallel transmission with equal minimum intervals and equal number of significant conditions on each channel, it is $m(1/T) \log_2 n$ $(m(1/T)$ in case of a two-condition modulation).
Note 2: In the above definition, the term 'parallel channels' is interpreted to mean: channels, each of which carries an integral part of an information unit, e.g. the parallel transmission of bits forming of character. In the case of a circuit comprising a number of channels, each of which carries information 'independently', with the sole purpose of increasing the traffic handling capacity, these channels are not to be regarded as parallel channels in the context of this definition.

date of manufacture
The date of issue of the document attesting that the individual aircraft or engine as appropriate conforms to the requirements of the type or the date of an analogous document.

datum line
A line fixed by the designer from which measurements are made when rigging or trueing the aircraft.

dead-beat indication
A term most commonly used with reference to instrument indications which give a minimum of oscillation.

dead reckoning   See NAVIGATION.

dead-stick landing
The landing of a powered aircraft with all engines inoperative.

decalage
The angle subtended between the chord lines of biplane wings, and between wings and stabiliser. It is positive when the upper surface (leading surface) is set at a greater angle of incidence than the other.

decametric
Having wave lengths in the order of 10.

Decca
A British hyperbolic radio navigation system in which the receiver translates signals from two or more synchronised stations into a continuous indication of ground positions.

decibel
One tenth of a bel (unit for comparing levels of power or sound). If P is the intensity of an actual sound, and p is the lowest audible intensity at that frequency, then the number of decibels (n) is calculated by the formula: $n = 10 \log P/p$.

decibel metre
The relative unit of power compared with a standard of 1 mW across an impedence of 600 ohms.

decision altitude   See DECISION HEIGHT.

decision height
A specified altitude or height in a precision approach at which a missed approach must be initiated if the required visual reference to continue the approach has not been established.
Note 1: Decision altitude (DA) is referenced to mean sea level (MSL) and decision height (DH) is referenced to the threshold elevation.
Note 2: The required visual reference means that section of the visual aids or of the approach area which should have been in view for sufficient time for the pilot to have made an assessment of the aircraft position and rate of change of position, in relation to the nominal flight path.

declared distance available
In relation to aerodromes:

(1) Takeoff run available (TORA). The length of runway which is available and suitable for the ground run of an aeroplane taking-off.
(2) Emergency distance (ED): the length of the declared takeoff run plus the length of stopway available.
(3) Takeoff distance available (TODA): the length of the declared takeoff run plus the length of clearway available.

(4) Landing distance available (LDA): the length of runway (or surface, when this is unpaved) which is available and suitable for the ground landing run of the aeroplane commencing at the landing threshold or displaced landing threshold.
(5) Accelerate-stop distance available (ASDA): The length of the takeoff run available plus the length of stopway available (if stopway is provided).

declared temperature
A temperature selected in such a way that when used for performance purposes, over a series of operations, the average level of safety is not less than would be obtained by using official forecast temperatures.

declination, magnetic
See MAGNETIC DECLINATION.

decode
The translation of various aeronautical codes, e.g. NOTAM, location indicators, Q-code, etc. into plain language.

decoder
The device used to decipher signals received from ATCRBS transponders to effect their display as select codes. See CODE and RADAR.

decometer
A phase meter used in the Decca navigation system; three are employed: purple, red and green.

decompression
A loss of pressure within a pressurised aircraft cabin.

dectrac
A compact display system integrated with Decca and designed for use in the smaller general aviation aircraft.

dedicated
Assigned for use in only one type of application.

dedicated runway
Normally a runway permanently dedicated to being the main instrument runway.

deepening
A decrease in the central pressure of a pressure system. Usually applied to a low rather than to a high, although, technically, it is acceptable in either sense.

deep stall
A stall condition normally associated with a rear-engined aircraft with a T-tail. It is associated with a flight condition where the angle of attack increases rapidly and the horizontal tailplane becomes inadequate for control in pitch.

defense visual flight rules (US)
A type of flight plan that must be filed by pilots planning VFR flight within or into an ADIZ (Air Defense Identification Zone), unless flying an aircraft at a true air speed of less than 180 knots.

definition
(Visual). The sharpness and clarity of display.

deflection
Bending or displacement of a structure.

degaussing
Surrounding an object with a coil carrying a decreasing a.c. current in order to demagnetise it.

degree Celsius (°C)
The special name for the unit kelvin for use in stating values of Celsius temperature.

degrees of freedom
The various rotations and translations that a moving body can do.

dehydration
The removal of chemically combined water from an organism or substance (drying merely removes excess water).

de-icing
Removal of ice accretion on aircraft. Normally conducted through the use of the following:
(1) Fluids. Certain fluids or paste when applied to the parts of an airframe make it difficult for ice to form. They can be applied either before flight or in certain installations which use fluid, during flight.
(2) Heating systems. Heater units can be used to prevent or reduce the possibility of ice formation on certain parts of the aircraft. Alternatively, electrical elements may be used to warm such specific areas as windshields, pitot tubes, propellers, etc.
(3) Rubber membranes. These take the form of leading edge coverings which pulsate, normally through the medium of the aircraft vacuum system. The membranes expand and crack the ice and therefore are only effective after ice has already formed.

delay indefinite (reason if known) expect approach further clearance (time)
Used by ATC to inform a pilot when an accurate estimate of the delay time and the reason for the delay cannot immediately be determined, e.g. a disabled aircraft on the runway, terminal or area saturation, weather below landing minimums. See also EXPECT FURTHER CLEARANCE.

delta aircraft
An aircraft with a delta wing planform which is triangular in shape and having a low aspect ratio.

delta-three angle (helicopter)
The coupling between blade flapping and pitch.

demodulation
Separation of signals from a modulated wave in radio, radar or TV.

density
Mass per unit volume, i.e. $kg.m^{-3}$.

density altitude
The altitude in the standard atmosphere corresponding to a particular value of air density. An aircraft will have the same performance characteristics as it would have in a standard atmosphere at this altitude.

density error
Airspeed indicators are calibrated to show the correct or true airspeed under conditions of the international standard atmosphere (ISA). However, ISA conditions will rarely occur in practice and because the density of air decreases with increase of altitude an error known as 'density error' occurs and has to be compensated for by the use of a navigation computer or other means when the true airspeed is required to be known, i.e. during navigation.

departure control
A function of an approach control facility providing air traffic control service for departing IFR and, under certain conditions, VFR aircraft. See also APPROACH CONTROL.

departure point
Aerodrome or point in space from which departure takes place.

departure procedures
ATC procedures established for aircraft which are departing from an aerodrome. Normally promulgated as Standard Instrument Departures (SIDs) and followed by aircraft during climbout to the minimum en-route altitude.

departure profile
The flight profile flown by an aircraft during departure from an aerodrome. The routes and heights used are normally constrained by such items as obstacle clearance, noise abatement, as well as vertical and horizontal clearance from other aircraft.

departure time
The time an aircraft becomes airborne.

depression
In meteorology, an area of low pressure; a low or trough. This is usually applied to a certain stage in the development of a tropical cyclone, to migratory lows and troughs, and to upper-level lows and troughs that are only weakly developed.

depth of section
The thickness of an aerofoil which varies in depth from the leading to trailing edge. At a given airspeed and angle of attack, the thicker the wing the greater the lift.

depth perception
The perception of spatial relationships, especially between objects in three dimensions.

derivative version
An aircraft gas turbine engine of the same generic family as an originally type-certificated engine and having features which retain the basic core engine and combustor design of the original model and for which other factors, as judged by the certificating authority, have not changed.

derived version of an aircraft
An aircraft which, from the point of view of airworthiness, is similar to the noise certificated prototype but incorporates changes in type design which may affect its noise characteristics.
Note: Where the certificating authority finds that the proposed change in design, configuration, power or mass is so extensive that a substantially new investigation of compliance with the applicable airworthiness regulations is required, the aircraft should be considered to be a new type design rather than a derived version.

designator
A number or letter code identifying a flight or an aerodrome.

design landing mass
The maximum mass of the aircraft at which, for structural design purposes, it is assumed that it will be planned to land.

design maximum weight
A specified weight which is used when stressing an aircraft structure for flight loads.

design takeoff mass
The maximum mass at which the aircraft, for structural design purposes, is assumed to be planned at the start of the takeoff run.

design taxying mass
The maximum mass of the aircraft at which structural provision is made for a load liable to occur during use of the aircraft on the ground prior to the start of takeoff.

destination alternate
An alternate aerodrome to which an aircraft may proceed should it become impossible or inadvisable to land at the aerodrome of intended landing.

detached shock wave
A shock wave proceeding ahead of the body which is causing it.

detector
An electronic circuit which de-modulates amplitude or pulse-modulated wave-forms.

detonation
Normal combustion within the cylinder occurs when the mixture ignites and burns progressively. This produces a normal pressure increase forcing the piston smoothly towards the bottom of the cylinder. There is, however, a limit to the amount of compression and degree of temperature rise that can be tolerated before the fuel/air mixture reaches a point at which it will ignite spontaneously without the aid of the sparking plug. This causes an instantaneous combustion, i.e. an explosion, and when this occurs an extremely high pressure is created within the

combustion area. This is called detonation and is very damaging to the piston top and valves.

deviation
(1) The error of a magnetic compass due to magnetic influences in the structure and equipment of an aircraft; the angular difference between the indicated bearing and the correct bearing of a radio signal received by radio direction finder, although the latter is more commonly known as 'quadrantal error'.
(2) A departure from a current clearance, such as an off-course manoeuvre to avoid weather or turbulence.
(3) Where specifically authorised and requested by the pilot, ATC may permit pilots to deviate from certain regulations.

deviation card
A card recording the deviations of a particular compass and indicating the compass direction corresponding to any desired magnetic direction.

dewpoint
The temperature at which the air is saturated with the water vapour it contains and condensation begins.

D factor
True altitude/pressure altitude.

DF fix
The geographical location of an aircraft obtained by one or more direction finders. See also DIRECTION FINDER.

DF guidance/DF steer
Headings provided to aircraft by facilities equipped with direction finding equipment. These headings, if followed, will lead the aircraft to a predetermined point such as the DF station or an airport. DF guidance is given to aircraft in distress or to other aircraft that request the service. Practice DF guidance is provided when workload permits. See also DIRECTION FINDER and DF FIX.

dielectric
Insulator, non-conductor of electrical current.

diesel engine
Internal combustion engine using high compression of oil/air mixture to reach ignition temperature of 540°C.

difference in depth of modulation
The percentage modulation depth of the larger signal minus the percentage modulation depth of the smaller signal, divided by 100.

differential aileron control
A mechanical device incorporated in the aileron system whereby the up-moving aileron will operate through a larger angle than the down-moving aileron.

differential positioning
Precise measurement of the relative positions of two receivers tracking the same GPS signal.

diffuser
A duct or chamber usually fitted with internal guide vanes which reduces the velocity of the fluid and increases static pressure. Associated with turbojet engines, superchangers.

diffusion of gases
All gases diffuse freely and distribute themselves equally within a container.

diffusion of light
The scattering of light rays through frosted glass, fog, etc.

digital
A system or device using discrete signals to represent particular values of a varying or fixed quantity numerically; a signal in such a system.

digital display
Numbers appearing on a screen instead of a pointer moving on a scale.

dihedral
The shape formed by two intersecting wings where the angle of the wings is above the horizontal.

dilution of precision
The multiplicative factor that modifies ranging error. Caused solely by the geometry between the user and his set of satellites. Commonly known as DOP or GDOP.

diode
Originally a thermionic valve with anode and cathode, nowadays a diode is a semiconductor with a P–N junction.

dip angle
The angle of a compass needle between the horizontal and the vertical effect of the magnetic pole's influence.

dipole
A radio aerial made of two rods.

direct
Straight-line flight between two navigational aids, fixes, points or any combination thereof. When used by pilots in describing off-airway routes, point defining direct route segments may become compulsory reporting points unless the aircraft is under radar contact.

direct access (computers)
Capability of reading data from a particular address in memory without having to access through a preceeding storage area.

direct current
The steady flow of electrons in one direction only.

direct entry
A specific type of entry to an instrument approach procedure or holding pattern.

direction

(1) Azimuth. The initial direction of the arc of a great circle; the angle between the plane of the great circle and the meridian of the place. As used in air navigation it is measured from north, in a clockwise direction, from 000° to 360°.

(2) Bearings. In air navigation, the same as azimuth. 'Azimuth' is preferred for use in celestial navigation; 'bearing' is preferred in all other forms of air navigation.

(3) Course (Track, in UK). The direction of the rhumb line, or the line of constant direction. As used in air navigation with the Lambert projection, it is measured at the meridian nearest halfway between the starting point and destination. 'Course' is also sometimes applied to the equisignal zone of (a) the localizer of an ILS, or (b) of one of the old 4-course L/M ranges.

(4) Heading. The horizontal direction in which the aircraft is pointed, in contradistinction to its path over the ground.

(5) Track. The rumb-line direction of the actual flight path of an aircraft over the ground. All the above directions may be true, magnetic or compass, according to whether they are referenced to true north, magnetic north or compass north.

(6) Wind. Always the true direction from which the wind blows. The one exception is that wind reported in landing instructions from a control tower is magnetic in direction.

directional aerial

An aerial which radiates or receives more efficiently in one particularly direction.

directional beacon

A transmitter which emits coded signals enabling a pilot to determine, by means of a compatible receiver, his bearing from the transmitter.

directional gyro (heading indicator)

An instrument which maintains for a limited time, by gyroscopic means, any direction for which it may be set.

direction finder

A radio receiver equipped with a directional sensing antenna used to take bearings on a radio transmitter. Specialised radio direction finders are used in aircraft as air navigation aids, others are ground-based primarily to obtain a 'fix' on a pilot requesting orientation assistance or to locate downed aircraft. A location fix is established by the intersection of two or more bearing lines plotted on a navigational chart using either two separately located direction finders to obtain a fix on an aircraft or by a pilot plotting the bearing indications of his DF on two separately located ground-based transmitters, both of which can be identified on his chart.

direction finder approach procedure

Used where DF is the sole aid and where another instrument approach procedure cannot be executed. DF guidance for an instrument approach is given by ATC facilities with DF capability.

direct transit area

A special area established in connection with an international airport approved by the authorities concerned and under their direct supervision for accommodation of traffic which is pausing briefly in its passage through the Contracting State.

direct transit arrangements
Special arrangements approved by the authorities concerned, by which traffic which is pausing briefly in its passage through the Contracting State may remain under their direct control.

discharge nozzle
A nozzle which discharges fuel into the cylinders of an engine, either directly or as part of a carburettor.

disc loading (helicopter)
Disc loading is comparable to wing loading in fixed wing aircraft. It is the ratio of the gross weight of the helicopter divided by the total area of the rotor disc. For example, if the disc area is $441.9\,\text{ft}^2$ and the fully loaded operational weight of the helicopter is 1670 lb, divide $441.9\,\text{ft}^2$ into 1670 lb. This will give the answer, in this case 3.77 lb per ft^2.

discontinuity
A zone with comparatively rapid transition of meteorological elements.

discrete beacon code
See DISCRETE CODE.

discrete code
As used in the air traffic control radar beacon system (ATCRBS), any one of the 4096 selectable mode 3/A aircraft transponder codes, except those ending in zero, e.g. 1201, 2317, 7777 are discrete codes; 0100, 1200, 7700 are non-discrete. Non-discrete codes are normally reserved for radar facilities that are not equipped with discrete decoding capability and for other purposes such as emergencies (7700). See also RADAR.

discrete frequency
A separate radio frequency for use in direct pilot–controller communications in air traffic control that reduces frequency congestion by controlling the number of aircraft operating on a particular frequency at one time. Discrete frequencies normally are designated for each control sector in en-route/terminal ATC facilities.

discrete signal
A signal characterised by being either 'on' or 'off'.

discriminator
A circuit which converts frequency or phase differences into amplitude variation.

disembarkation
The leaving of an aircraft after a landing, except by crew or passengers continuing on the next stage of the same through-flight.

dish
A reflector for centimetric radar waves, whose surface forms part of a paraboloid or sphere.

disorientation (physical senses)
A state which is primarily produced by receiving conflicting messages through the three main sensory systems, i.e. visual, or motion sensing from the inner ear, and

position sensing involving nerves in the skin muscles and joints of the human body.

displaced threshold
A threshold that is located at a point on the runway other than the designated beginning of the runway. See also THRESHOLD.

displacement
Movement measured in terms of distance and direction.

displacement error
The angular or linear displacement of any point of zero DDM with respect to the nominal course line or the nominal ILS glide path respectively.

displacement sensitivity (localiser)
The ratio of measured DDM to the corresponding lateral displacement from the appropriate reference line.

display
A visual presentation of data in a manner which permits interpretation by a controller.

disposable load [UK]
The weight of all persons and items of load, including fuel and other consumable fluids, carried in the aircraft, other than the basic equipment and variable load.

distance marker
(1) Numerals or marks painted on an instrument runway to indicate distance to the upwind end of the runway.
(2) Range marker used on a radar display.

distance measuring equipment
(1) Equipment (airborne and ground) used to measure, in nautical miles, the slant range distance of an aircraft from the DME navigational aid. See also TACAN and VORTAC.
(2) UHF radio aid that determines the distance of an airborne interrogator from a ground transponder, by measuring the time of transmission to and from the transponder.

distress and diversion section
An ATC emergency service for radio-equipped aircraft in flight. It provides various forms of assistance when contacted on the aeronautical emergency frequency.

distress phase
A situation wherein there is reasonable certainty that an aircraft and its occupants are threatened by grave and imminent danger or require immediate assistance.

disturbance
In meteorology, applied rather loosely:

(1) any low pressure or cyclone, but usually one that is relatively small in size.
(2) an area where weather, wind, pressure etc. show signs of cyclonic development.

(3) any deviation in flow or pressure that is associated with a disturbed state of the weather, i.e. cloudiness and precipitation;

(4) any individual circulatory system within the primary circulation of the atmosphere.

disturbing moment
A moment which tends to upset or rotate an aircraft about its axes.

ditching
The forced landing of an aircraft on water.

ditching drill
Procedures to be performed by the crew and passengers of an aircraft in the event of ditching.

diurnal
Daily, especially pertaining to a cycle completed within a 24-hour period, and which recurs every 24 hours.

dive brakes
See SPEED BRAKES.

divergence
The condition that exists when the distribution of winds within a given area is such that there is a net horizontal flow of air outward from the region. In divergence at lower levels, the resulting deficit is compensated for by subsidence of air from aloft; consequently the air is heated and the relative humidity lowered, making divergence a warming and drying process. Low-level divergent regions are areas unfavourable to the occurrence of clouds and precipitation. The opposite of convergence.

divergent
Going away in different directions.

diversion
The act of proceeding to an aerodrome other than one at which a landing was intended.

DME dead time
A period of blanking in the receiver of the transponder to protect it during reply transmission and to prevent responses due to echoes resulting from multipath effects.

DME distance
The line of sight distance (slant range) from the source of a DME signal to the receiving antenna.

DME fix
A geographical position determined by reference to a navigational aid that provides distance and azimuth information. It is defined by a specific distance in nautical miles and a radical or course (i.e. localiser) in degrees magnetic from that aid. See DISTANCE MEASURING EQUIPMENT and FIX.

DME/P
The distance measuring element of the MLS, where the 'P' stands for precise distance measurement.

DME separation
Spacing of aircraft in terms of distances (nautical miles) determined by reference to distance measuring equipment (DME). see also DISTANCE MEASURING EQUIPMENT.

DME/W
Distance measuring equipment, primarily serving operational needs of en-route or TMA navigation, where the 'W' stands for wide spectrum characteristics.

document review into video entry
A method of storing information on aircraft servicing into a central bank of access information.

Dod flip [US]
Department of Defense Flight Information Publications used for flight planning, en-route and terminal operations. Flip is produced by the Defense Mapping Agency for world-wide use. United States Government Flight Information Publications (en-route charts and instrument approach procedure charts) are incorporated in DOD flip for use in the national airspace system (NAS).

dog-leg
A deliberate manoeuvre to diverge from track and to return back again further along the intended track.

doldrums
The equatorial belt of calm or light and variable winds between the two trade wind belts.

domestic flight
A flight made within the geographic boundaries of a State.

Doppler effect
The way in which the frequency appears to increase as the wave source approaches and decreases as it goes away; the high pitch sound of an approaching aircraft and the low pitched sound as it goes away. This effect is used in radar to determine the velocity of moving objects.

Doppler navigation system
A dead reckoning system consisting of Doppler radar and a computer which, with heading information from a compass, calculates the position of the aircraft.

Doppler radar
A primary radar system which utilises the Doppler effect to measure two or more of ground speed, drift angle, longitudinal velocity, lateral velocity and vertical velocity.

Doppler shift
(1) The difference between transmit and receive frequencies in a system subject to the Doppler effect.

(2) The apparent change in the frequency of a signal caused by the relative motion of the transmitter and receiver.

Doppler spectrum
A band of Doppler shift frequencies produced by a Doppler radar with a finite beam width.

double channel simplex
Simplex using two frequency channels, one in each direction.
Note: This method was sometimes referred to as cross-band.

downdraft
A relatively small-scale downward current of air; often observed on the lee side of large objects restricting the smooth flow of the air or in precipitation areas in or near cumuliform clouds.

down time
The time period an aircraft is on the ground. Also used to indicate the time of landing.

downwash
The deflected air flow after passing over a wing.

downwind leg
See TRAFFIC PATTERN.

drag
The resistance to motion along the line of flight.

drag chute
A parachute device installed on certain aircraft that is deployed on landing roll to assist in deceleration of the aircraft.

drag coefficient
A coefficient of the amount of drag experienced by a body.

drag hinge (helicopter)
A hinge which permits the rotor blade to pivot front and rear in the plane of rotation.

drift angle
The angle between the heading of an aircraft and its track, or flight path over the ground.

drift correction
The angular correction to the track being made to return to the original planned track.

drift sight
An instrument for determining the angle of drift, often accomplished by observing the apparent motion of points on the earth's surface along a grid incorporated in the instrument.

drifting snow
A type of hydrometeor composed of snow particles picked up from the surface, but carried to a height of less than 6 ft.

drizzle
A form of precipitation. Very small water drops that appear to float with the air currents while falling in an irregular path (unlike rain, which falls in a comparatively straight path, and unlike fog droplets which remain suspended in the air).

drooping ailerons
Ailerons which are designed to droop normally 10° to 15° when the flaps are lowered, thus improving lateral control.

droop leading edge
The leading edge of a wing which is hinged in order to rotate down and provide more lift at slow speed, i.e. during the takeoff and landing phase.

dropsonde
A radiosonde dropped by parachute from an aircraft to obtain soundings (measurements) of the atmosphere below.

dry adiabatic lapse rate
The rate of decrease of temperature with height when unsaturated air is lifted adiabatically (due to expansion as it is lifted to lower pressure). See also ADIABATIC PROCESS.

dry bulb
A name given to an ordinary thermometer used to determine temperature of the air; also used as a contraction for dry-bulb temperature. Compare WET BULB.

dry bulb temperature
The temperature of the air.

dry cell
A small battery cell (leclanché) containing no freely-flowing liquids; used in calculators, etc.

dry ice (CO_2)
Solid carbon dioxide, melting point −78.5°C.

dry lease
An arrangement whereby an aircraft operator leases an aircraft without flight or cabin crew or supporting services, e.g. fuel, etc.

dry sump
An engine lubrication system where oil does not remain in the crankcase but is pumped out as fast as it flows in, and is returned to a separate reservoir.

dual instruction time
Flight time during which a person is receiving flight instruction from a properly authorised pilot on board the aircraft.

ducted fan (or propeller)
A fan or propeller driven by the engine and operating in a duct to improve propulsive efficiency.

ductility
Capacity of being drawn out into a wire.

duplex
A method in which telecommunications between two stations can take place in both directions simultaneously.

duplexer
A device which permits sharing of one circuit or transmission channel by two signals, in particular use of one antenna for reception and transmission.

Duplicate inspection
Defined as an inspection first made and certified by one qualified person and subsequently made and certified by a second qualified person.

dust
A type of lithometeor composed of small earthen particles suspended in the atmosphere.

dust devil
A small, vigorous whirlwind, usually of short duration, rendered visible by dust, sand and debris picked up from the ground.

duster
As dust storm.

duty
In relation to a flightcrew member, any continuous period during which the crew member is required to conduct any task associated with the business of an aircraft operation.

duty runway   See RUNWAY IN USE.

dynamic heating
The heating of aircraft surfaces by motion through the air. Generation of heat occurs when air comes to rest against the leading edges of the aircraft or by the action of viscosity as the air flows over the aircraft surfaces.

dynamic RAM
A type of random access memory (RAM) in which data stored will fade unless periodically renewed.

dynamics
A study of moving bodies and the forces acting upon them.

dynamic sidelobe level
The level that is exceeded 3% of the time by the scanning antenna far-field radiation pattern exclusive of the main beam as measured at the function scan rate using a 26 kHz beam envelope video filter. The 3% level is determined by the ratio of the sidelobe duration which exceeds the specified level to the total scan duration.

dynamic stability
The characteristics of an aircraft following a disturbance from its state of equilibrium.

dynamic thrust
The work done by a propeller or fan by imparting a forward motion

dynamo
Converts mechanical energy into electrical current by rotating an armature in a magnetic field-generator.

dyne
A unit of force which will cause a mass of 1 gramme to accelerate at 1 cm per second, per second $= 10^{-5}$ newtons.

Dzus fastener
Used to secure moveable or hinged panels to the airframe. Consists of a countersunk type of screw with a slotted head and shank to lock onto a wire anchor fixed to the airframe.

E

earth
Polar diameter = 12794 km, 6860 NM; equatorial diameter = 12757 km, 6884 NM mass = 5.796 × 10²⁴kg; polar circumference = 39935 km, 21550 n miles; equatorial circumference = 40077 km, 21626 n miles; shape: oblate spheroid.

earthing
Earthing, or grounding as it is often termed, refers to the return of electrical current to the conducting mass of the earth, or ground itself. If considered as a single body, the earth is so large that any transfer of electrons between it and another body fails to produce any measurable change in its state of electrification. Therefore it can be regarded as electrically neutral.

earth's crust (lithosphere)
A few miles thick. Composition by weight = 47% oxygen, 28% silicon, 8% aluminium, 4.5% iron, 3.5% calcium, 2.5% sodium, 2.5% potassium, 2.2% magnesium, 0.5% titanium, 0.2% hydrogen, 0.2% carbon, 0.1% sulphur and 0.1% phosphorus.

earth station
A location on an aircraft where a negative connection from an electrical circuit can be coupled and thus act as a negative busbar.

earth stations
Components of the INMARSAT satellite systems providing communications and navigation facilities. Coast, earth and aeronautical ground stations provide an interface between the space segment of the system and the national and international telecommunications networks. Mobile earth stations are the satellite communications terminals which are installed on individual ships, aircraft or other vehicles.

echo
Sound waves reflected from a solid object or other reflecting substance; in radar:
(1) the energy reflected or scattered by a target;
(2) the radar scope presentation of the return from a target.

eclipse
When the shadow of the earth falls on the moon, or when the shadow of the moon falls on the earth.

eddy
A whirl or circling current of air or water, moving contrary to the general flow.

eddy currents
Undesirable induced electric currents occurring at the cores of electromagnets (armatures, transformers etc). Produce heat and waste energy.

effective acceptance bandwidth
The range of frequencies with respect to the assigned frequency for which reception is assured when all receiver tolerances have been taken into account.

effective adjacent channel rejection
The rejection that is obtained at the appropriate adjacent channel frequency when all relevant receiver tolerances have been taken into account.

effective coverage
The area surrounding an NDB within which bearings can be obtained with an accuracy sufficient for the nature of the operation concerned.

effective horsepower
Power delivered to the propeller.

effective intensity
The effective intensity of a flashing light is equal to the intensity of a fixed light of the same colour which will produce the same visual range under identical conditions of observation.

effective margin
That margin of an individual apparatus which could be measured under actual operating conditions.

effective pitch
The distance the aircraft travels along its flight path in one revolution of the propeller.

effective sidelobe level
That level of scanning beam sidelobe which in a specified multipath environment results in a particular guidance angle error.

efficiency
The output energy divided by the input energy expressed as a percentage.

Einstein's equation
$E = mc^2$ showing that the energy equivalent of a mass is equal to the mass multiplied by the square of the velocity of light ($c = 299792.5$ kilometres per second).

ejection seat
A seat that can be ejected from an aircraft, carrying the pilot with it, in an emergency.

elastic modulus
The ratio of stress to strain. The stress is measured in N/m^2. The resulting strain may be elongation, twist or shear or a change in volume (bulk). The elastic limit is the amount of stress which, if exceeded, will permanently change the shape of the body.

electric charge
The quantity of electricity on an electrically charged body, or passing at a point in an electric circuit during a given time.

electric energy
Measured in joules (J) which are equivalent to watt seconds, newton metres, (0.238846 calories). 10^7 ergs, 0.737561 lb.
(1 electron volt = 1.602192×10^{-19} joules). One kilowatt — hour equals 3 600 000 joules or 859845 calories.

electric potential
Measured by the energy of a unit positive charge at a point, expressed relative to zero potential, or earth.

electric power
Measured in watts, the product of volts × amps.

electro cardiogram
A device which records electrical changes within the heart muscle and can be used to diagnose abnormalities.

electrode
A conductor through which a current enters or leaves an electrolyte. The positive electrode is called the anode and the negative one is called the cathode; a component that emits (cathode) or collects (anode) electrons or ions in a semiconductor device.

electrodynamometer
Used for measuring volts, amps and watts, both a.c. and d.c.

electroluminescent
A property of devices which convert electrical energy to light.

electrolysis
Passing an electric current through a liquid to decompose it by ionisation.

electrolyte
A liquid which is chemically changed by the current it conducts.

electromagnet
A magnet made by passing a current through a coil wound around a soft iron core.

electromotive force
Measured in volts which are defined as Joules per coulomb.

electron
A subatomic particle with a negative charge (its anti-particle is the positively charged positron). Its mass is 9.109558×10^{-31} kg and its diameter is 5.63554×10^{-15} m. A free electron is one which is not confined to an atomic orbital.

electron gun
In a cathode ray tube, comprising an emitter (cathode), an aperture (anode), together with additional electrodes to focus the beam.

electronic interference
Disturbance causing interference in electronic equipment.

electroplating
The object to be plated forms the cathode, i.e. it is connected to the negative. The plating metal is contained in the electrolyte in the form of a dissolved salt.

electrostatic capacity
Ability to hold an electrical charge.

element
A substance in which all the atoms have the same atomic number.

elevation
The vertical height above sea level, or a point on the earth's surface.

elevator
A horizontal control surface hinged to the rear of the tailplane.

eleven-year period
The period of time through which sunspots (which affect the weather) regularly increase and decrease.

elevons
A combination of ailerons and elevators, as on delta wing aircraft.

elinvar
A type of steel which does not expand or contract with changes in temperature.

embarkation
The boarding of an aircraft for the purpose of commencing a flight, except by such crew or passengers as have embarked on a previous stage of the same through-flight.

emergency
A distress or urgency condition.

emergency ceiling
The highest altitude at which a multi-engined aircraft can maintain a rate of climb of 50 ft per minute with one engine inoperative.

emergency descent
A rapid descent from operating altitude due to an in-flight emergency.

emergency distance
The length of the declared takeoff run plus the length of stopway available.

emergency locator transmitter
A radio transmitter attached to the aircraft structure that operates from its own power source on 121.5 MHz and 243.0 MHz. It aids in locating downed aircraft by radiating an audio tone, two or four times per second. It is designed to function without human action after an accident.

emergency phase
A generic term meaning, as the case may be, uncertainty phase, alert phase or distress phase.

emergency safe altitude   See MINIMUM SAFE ALTITUDE.

emery
A mixture of a corundum (Al_2O_3) and magnetite (Fe_3O_4) used as an abrasive.

empennage
The entire rear section of an aircraft including all fixed and movable tail surfaces.

encoding altimeter
A pneumatic altimeter with a parallel coded output of 9 to 11 bits representing the aircraft's height above mean sea level to the nearest 100 ft. It operates in conjunction with an aircraft transponder and passes altitude information to ATC.

endothermic
The absorption of heat.

endurance
The maximum time an aircraft can fly on the available fuel.

energy
The capacity to do work. Its different forms are: heat, chemical, electrical, potential, kinetic and nuclear; all occur in the presence of matter and are interchangeable. Only radiant energy can exist without matter. Einstein's equation $E = mc^2$ (proved by the increase in mass of particles in accelerators) shows that energy can be converted into mass, and vice versa.

engine cowling
Hinged or removable coverings round an aircraft engine.

engine icing
A situation which can produce a significant loss of power is the ingestion of large water droplets into the air intakes of carburettors. This can affect the fuel/air mixture ratio, and cause power loss, but when temperatures in the engine induction system are at or below 0°C, these water droplets can freeze on the internal surfaces of the system and form carburettor ice. This type of icing is caused during moderate to high humidity conditions in conjunction with the temperature drop due to the evaporation of fuel and the expansion of air as it passes through the carburettor. In small aero engines this temperature drop can be as much as 20°C. Water vapour contained in the air is condensed by the cooling and if the temperature in the carburettor reaches 0°C or below, moisture is deposited as frost or ice on the inside of the carburettor passages and the throttle butterfly valve. Even a slight accumulation of this deposit will reduce power and in larger amounts may lead to complete engine stoppage.

engine mounting
The structure by which an engine is attached to the airframe.

engine rating
The power permitted under different conditions, i.e. maximum takeoff, maximum continuous or weak mixture.

en-route air traffic control service [US]
Air traffic control service provided for aircraft on an IFR flight plan, generally by centres, when these aircraft are operating between departure and destination terminal areas. When equipment capabilities and controller workload permit, certain advisory/assistance services may be provided to VFR aircraft. See also AIR ROUTE TRAFFIC CONTROL CENTRE.

en-route alternate
An aerodrome at which an aircraft would be able to land after experiencing an abnormal or emergency condition while en-route.

en-route automated radar tracking system [US]
An automated radar and radar beacon tracking system (EARTS). Its functional capabilities and design are essentially the same as the terminal ARTS IIIA system, except that the EARTS is capable of employing both short-range (ASR) and long range (ARSR) radars, uses full digital radar displays and has fail-safe design. See also AUTOMATED RADAR TERMINAL SYSTEMS.

en-route chart   See AERONAUTICAL CHART.

en-route clearance
A clearance covering the flight path of an aircraft after takeoff to the point at which an approach to land is expected to commence.
Note: In some circumstances it may be necessary to subdivide this clearance, e.g. into sections divided by control area boundaries or into the departure, climb or descent phases of flight.

en-route climb
The climb to altitude or flight level on the desired track.

en-route descent
Descent from the en-route cruising altitude that takes place along the route of flight.

en-route fight advisory service [US]
A service specifically designed to provide, upon pilot request, timely weather information pertinent to the type of flight, intended route of flight and altitude. The Flight Service Stations (FSS) providing this service is listed in the Airport/Facility Directory. See also FLIGHT WATCH.

en-route minimum safe altitude warning [US]
A function of the NAS stage. An en-route computer that aids the controller by alerting him when a tracked aircraft is below or predicted by the computer to go below a predetermined minimum IFR altitude (MIA).

entry fix
The first reporting point, determined by reference to a navigation aid, over which an aircraft passes or is expected to pass upon entering a flight information region or a control area.

environmental lapse rate
The temperature lapse rate of the environmental air which surrounds air moving vertically.

ephemeris
The predictions of current satellite positions that are transmitted to the user in the data message.

equal area projection
A projection in which a square inch in one part of the map represents the same area on the earth's surface as a square inch in any other part of the map.

equator
(1) Magnetic. An imaginary line on the earth's surface approximately equidistant from the magnetic poles.
(2) Terrestrial. An imaginary line on the earth's surface traced by a radius at right angles to the mid-point of the axis.

equilibrium
During flight, an aircraft is subject to the effects of the forces of lift, weight, thrust and drag. In level flight at a steady air speed, the aircraft will be in equilibrium, i.e. the lift will equal the aircraft weight, and thrust will equal drag.

equinox
Vernal, 21 March; autumnal, 21 September. The two days of the solar year on which all latitudes receive 12 hours daylight and 12 hours darkness.

equisignal zone
A narrow zone within which two transmitted signals overlap to provide signals of equal strength (the 'localizer' beam) along the centreline of an ILS runway. This affords lateral guidance for an instrument approach.

equivalence (charts)
The presentation of area in correct proportion to that on the earth.

equivalent air speed   See AIR SPEED.

equivalent isotropically radiated power
The product of the power supplied to the antenna and the antenna gain in a given direction relative to an isotropic antenna (absolute or isotropic gain).

erg
Work done by a force of 1 dyne over a distance of 1 centimetre = 10^{-7} joules.

escape velocity
The speed a body must attain to escape from a gravitational field:

(1) From the earth's surface = 11,200 m/s or about 21,710 kts.
(2) From the moon's surface = 2,370 m/s or about 4,600 kts.

estimated ceiling
A ceiling classification applied when the ceiling height has been estimated by the observer or has been determined by some other method, but when because of the

specified limits of time, distance or precipitation conditions, a more descriptive classification cannot be applied.

estimated elapsed time
The estimated time required to proceed from one significant point to another.

estimated off-block time.
The estimated time at which the aircraft will commence movement associated with departure.

estimated time of arrival
For IFR flights, the time at which it is estimated that the aircraft will arrive over that designated point, defined by reference to navigation aids from which it is intended that an instrument approach procedure will be commenced or, if no navigation aid is associated with the aerodrome, the time at which the aircraft will arrive over the aerodrome. For VFR flights, the time at which it is estimated that the aircraft will arrive over the aerodrome.

European Air Navigation Planning Group [ICAO]
Set up by the Council of ICAO in 1972. The objectives of the Group are to ensure the continuous and coherent development of the European Regional Plan as a whole and in relation to that of adjacent regions; to identify specific problems in the air navigation field and propose resolving actions. In conducting its task the Group creates sub groups of experts to give advice on specific problems.

European Association of Aerospace Manufacturers
Formed by a number of aerospace manufacturer's trade associations in the European region. The objective of the Association is to promote the development of the European airspace industry by making it more competitive as a whole and by trying to create, for its benefit, a domestic European market. To meet this objective the Association studies all problems linked with this development, aiming to find solutions and define co-ordinated plans of action for their implementation. The Association also represents its members with regard to relevant authorities, other European associations and institutions, and international organisations.

European Business Aviation Association
Formerly International Business Aircraft Association (Europe), the Association helped to form the International Business Aviation Council and represents it officially with observer status in EUROCONTROL's discussions with users of the European airspace. The organisation combines the interests of corporate business aircraft with those of the charter/taxi business aircraft operations. Maintains contact with national and international bodies and participates in ECAC committees.

European Civil Aviation Conference
An organisation formed by a number of European States (23 in 1989) with the following objectives and functions:

(1) To review generally the development of European air transport in order to promote the co-ordination, the better utilisation and the orderly development of such air transport;
(2) To consider any special problem that may arise in this field.

The conference brings within its scope all matters relevant to these objects, taking into account the following principles regarding the selection of items for the work programme:

(1) Importance and interest of the subject to a large number of Member States or to other European organisations;
(2) Possibility of an acceptable solution to the problems involved;
(3) Peculiarity of the problem to Europe or possibility of making an effective contribution to the normal work of ICAO.

The functions of the Conference are consultative and its resolutions, recommendations or other conclusions are subject to the approval of governments. The Conference is composed of the States invited to be members of the 1954 Strasbourg Conference on Co-ordination of Air Transport in Europe, together with such other European States as the Conference may unanimously admit as members. All Member States have an equal right to be represented at the sessions and meetings of the Conference. The Conference maintains close liaison with ICAO in order, through regional co-operation, to help achieve the aims and objectives of that Organisation. It avails itself of the services of the ICAO Secretariat, and establishes relations with other governmental or non-governmental international organisations concerned with European air transport.

European General Aviation Safety Foundation
An organisation formed in 1989 by the Aircraft Owner and Pilot Associations in the European region. It is a non-profit, totally independent organisation which promotes pilot proficiency and safety through quality training, education, research, analysis and dissemination of safety information. The Foundation's objectives are to:

(1) Conduct research to identify safety trends and principal causes of accidents in European General Aviation and, in concert with national AOPAs and other organisations, communicate this information to the general aviation community via the most cost effective means available.
(2) Acquire, develop, implement and conduct general aviation educational and training programmes aimed at reducing the number of accidents and incidents and enhance the level of general aviation safety.
(3) Develop the quality of its programmes and broaden the scope of its services in order to establish EGASF as the recognised and respected safety organisation for general aviation throughout Europe.

European Helicopter Association
Representative body for European helicopter associations aiming to standardise regulatory and technical aspects of helicopter operations.

European Organisation for Civil Aviation Electronics
EUROCAE was formed in 1963 as a non-profit making organisation to provide a European forum for collaboration between administrations responsible for civil aviation, designers, manufacturers and users of electronic equipment for civil aviation; and to communicate with other interested national and international organisations.

European Organisation for the Safety of Air Navigation
This organisation, more generally known as EUROCONTROL, was established by an International Convention which was signed on 13 December 1960 and ratified by the Federal Republic of Germany, Belgium, France, the United Kingdom, Luxembourg, the Netherlands and Ireland. The Convention came into force on 1 March 1963 for an initial period of 20 years. The object was to create an air traffic control system which would be independent of national boundaries. Currently EUROCONTROL carries out numerous functions in relation to air traffic control. The original Convention laid down that air traffic control in the upper airspace which in most cases is the airspace above 25 000 ft was to be provided by EUROCONTROL. Under the amended Convention, this obligation is changed to the possibility of the organisation providing air traffic services for the whole airspace (including lower airspace) on behalf of one or more Member States at their request. The tasks assigned to EUROCONTROL under the amended Convention entail complex technical and operational activities in such areas as operational research, the compiling and utilisation of traffic statistics, engineering, the commissioning of highly automated ATC systems and the associated software, and the study of existing and future air traffic control systems and the evaluation of their component parts. In addition the Member States and non-Member States may entrust the Organisation with specific tasks such as to assist in the design and setting-up of, and to operate, air traffic facilities, and to assist in the calculation and collection of route charges. The EUROCONTROL Organisation consists of two organs:

(1) The Permanent Commission for the Safety of Air Navigation. The policy making body.
(2) The Agency for the Safety of Air Navigation. An executive body administrated by a Committee of Management and a Director General.

European Regional Airline Organisation
Formed to identify, protect and promote the interests of all aspects of regional air transport in Europe.

Eustachian tube
The narrow duct leading from the middle ear to open into the back of the throat. By this means the middle ear communicates with the outside air so that, normally, any pressure difference between the middle ear and the surrounding atmosphere is equalised.

evaporation
The change from liquid to vapour (gas), requiring the latent heat of vaporisation. This can take place at any temperature at which the fastest moving molecules reach sufficient kinetic energy to escape from the surface.

evaporation icing
Ice formation in an engine induction system as a result of cooling by evaporation.

excess thrust horse power
Difference between the thrust horse power available and the horse power required. This determines the rate of climb of an aircraft.

execute missed approach
Instructions issued to a pilot making an instrument approach, meaning to continue inbound to the missed approach point and execute the missed approach procedure as described on the instrument approach procedure chart, or as previously assigned by ATC. The pilot may climb immediately to the altitude specified in the missed approach procedure upon making a missed approach. When conducting an ASR or PAR approach, execute the assigned missed approach procedure immediately upon receiving instructions to 'execute missed approach'.

exemption
An authorisation issued by an appropriate national authority providing relief from the national or international regulations.

exhaust gas analyser
An electrical instrument giving an indication as to the proportion of carbon monoxide in exhaust gases.

exhaust gas temperature gauge
An instrument which measures the temperature used for control of the air/fuel ratio within efficient limits.

exit fix
The last reporting point, determined by reference to a navigation aid, over which an aircraft passes, or is expected to pass, before leaving a flight information region or a control area.

exosphere
Lies beyond the ionosphere starting at 400 km amsl. Exosphere molecules have a 50% chance of escape into space.

expansion of gases
According to Charles's law the volume of a perfect gas changes by 1/273 of its volume at 0°C, with each change in its temperature of 1°C (provided the pressure remains constant).

expect (altitude) at (time) or (fix)
Used under certain conditions in a departure clearance to provide a pilot with an altitude to be used in the event of two-way communications failure.

expect departure clearance time [US]
Used in fuel advisory departure (FAD) programme. The time the operator can expect a gate release. Excluding long distance flight, an EDCT will always be assigned even though it may be the same as the estimated time of departure. The EDTC is calculated by adding the ground delay factor. See also FUEL ADVISORY DEPARTURE.

expected
Used in relation to various aspects of performance (e.g. rate or gradient of climb), this term signifies the standard performance for the type, in the relevant conditions (e.g. mass, altitude and temperature).

expected approach clearance time
The time at which it is expected that an aircraft will be cleared to commence an approach to landing.

expected approach time
The time at which ATC expects that an arriving aircraft, following a delay, will leave the holding point to complete its approach for a landing.
Note: The actual time of leaving the holding point will depend upon the approach clearance.

expect further clearance (time)
The time a pilot can expect to receive clearance beyond a clearance limit.

expect further clearance via (airways, route or fixes)
Used to inform a pilot of the routeing he can expect if any part of the route beyond a short-range clearance limit differs from that filed.

expedite
Used by ATC when prompt compliance is required to avoid the development of an imminent situation.

experimental mean pitch
The distance through which a propeller advances along its axis during one revolution.

explosive article
An article containing one or more explosive substances.

explosive decompression
A sudden and marked reduction in pressure caused by a disastrous leak in a pressurised cabin.

explosive substance
A solid or liquid substance (or a mixture of substances) which is in itself capable by chemical reaction of producing gas at such a temperature and pressure and at such a speed as to cause damage to the surroundings. Included are pyrotechnic substances even when they do not evolve gases. A substance which is not itself an explosive but which can form an explosive atmosphere of gas, vapour or dust is not included.

extended centreline
An imaginary line extending along the runway centreline into the area beyond either end of the runway.

extended over-water operation
(1) With respect to aircraft other than helicopters. An operation over water at a horizontal distance of more than 50 nautical miles from the nearest shoreline.

(2) With respect to helicopters. An operation over water at a horizontal distance of more than 50 nautical miles from the nearest shoreline and more than 50 nautical miles from an offshore heliport structure.

extended range operation
Any flight by an aeroplane with two turbine power-units where from any point on the route the flight time at the one power-unit inoperative cruise speed to an adequate aerodrome is greater than the threshold time approved by the State of the operator.

external load
A load that is carried, or extends, outside of the aircraft fuselage.

external-load attaching means
The structural components used to attach an external load to an aircraft, including external-load containers, the backup structure at the attachment points and any quick-release device used to jettison the external load.

extratropical low
Any cyclone that is not a tropical cyclone, usually referring to the migratory frontal cyclones of middle and high latitudes. Sometimes called extratropical cyclone, extratropical storm.

extremely high frequencies
Radio frequencies from 30 000 to 300 000 MHz.

eye
The roughly circular area of calm or relatively light winds and comparatively fair weather at the centre of a well-developed tropical cyclone. A wall cloud marks the outer boundary of the eye.

eyebrow panel
A panel of instruments, controls, etc. fitted in the cabin roof above and behind the windscreen.

F

facilitation
The provision of facilities relating to the departure and arrival of passengers and aircraft at airports.

facility availability
The ratio of actual operating time to specified operating time.

facility failure
Any unanticipated occurrence which gives rise to an operationally significant period during which a facility does not provide service within the specified tolerances.

facility performance category I–ILS [ICAO]
An ILS which provides guidance information from the coverage limit of the ILS to the point at which the localiser course line intersects the ILS glide path at a height of 60 m (200 ft) or less above the horizontal plane containing the threshold.
Note: This definition is not intended to preclude the use of facility performance category I–ILS below the height of 60 m (200 ft), with visual reference where the quality of the guidance provided permits, and where satisfactory operational procedures have been established.

facility performance category II–ILS [ICAO]
An ILS which provides guidance information from the coverage limit of the ILS to the point at which the localiser course line intersects the ILS glide path at a height of 15 m (50 ft) or less above the horizontal plane containing the threshold.

facility performance category III–ILS [ICAO]
An ILS which, with the aid of ancillary equipment where necessary, provides guidance information from the coverage limit of the facility to, and along the surface of, the runway.

facility reliability
The probability that the ground installation operates within the specified tolerances.

factored field lengths
Takeoff and landing distances which have been increased to take account of surface conditions, wind variation and other factors which may affect the performance of the aircraft.

factor of safety
A design factor used to provide for the possibility of loads greater than those assumed, and for uncertainties in design and fabrication.

faded
A radar expression meaning the signal has disappeared from the screen.

Fahrenheit
Scale of temperature with ice melting at 32° and water boiling at 212°. Therefore, conversion to Celsius is by $(\text{Temp}°F - 32) \times \frac{5}{9} = \text{Temp}°C$.

fail-safe
A system in which no crack can cause complete failure in a structure, and any damage which occurs can be detected before it becomes catastrophic.

fairing
Additions to any structure to reduce drag.

fall-out
Radioactive particles which fall to the earth's surface after a nuclear explosion as follows:

(1) Local fall-out of large particles from the fireball.
(2) Tropospheric fall-out of fine particles spread around that latitude by winds.
(3) Stratospheric fall out of fine particles spread up into the stratosphere. Over a period of years these are spread over the whole surface of the earth.

fall wind
A wind blowing down a mountainside; or any wind having a strong downward component. Fall winds include the foehn, mistral, bora etc.

false cirrus
Cirrus-like clouds at the summit of a thunder cloud; more appropriately called 'thunderstorm cirrus'.

false horizon
A sloping cloud formation, an obscured horizon, a dark scene spread with ground lights and stars, and certain geometric patterns of ground lights can provide inaccurate visual information for aligning the aircraft correctly with the actual horizon. The disorientated pilot may place the aircraft in a dangerous attitude.

false start
An attempt to start an engine when the fuel fails to ignite.

fan marker
A VHF radio aid having a fan-shaped radiation pattern and located along the localiser beam of an ILS to provide a position fix and information as to the distance still to go. On the surface transmitting vertically in either a 'bone' or 'elliptical' pattern on 75 MHz, and received by aircraft only when within the transmission pattern. May also be located away from an ILS, e.g. on an airway.

farad
A unit of capacitance measured as a p.d. of 1 volt when the plates are charged with 1 coulomb. Most capacitance is measured in μf (microfarads), i.e. $F \times 10^{-6}$.

fast erect
The application of a larger voltage than required for normal running to a gyro in order to reduce the time taken to achieve operating speed.

fast file [US]
A system whereby a pilot files a flight plan via a telephone that is tape recorded and then transcribed for transmission to the appropriate air traffic facility. Locations having a fast file capability are contained in the *Airport/Facility Directory*.

fast-switching channel
A single channel which rapidly samples a number of satellite ranges. 'Fast' means that the switching time is sufficiently fast (2–5 milliseconds) to recover the data message.

fathom
Unit of depth, 6 ft, used on some marine maps.

fatigue life
The minimum time, expressed in hours or load cycles, that a structure is designed for use without fatigue failure.

fatigue of metals
Weakness in metals because of changes in the crystalline structure caused by repeated stresses.

fatigue strength
The maximum stress or loading that can be applied for a specified number of cycles without failure of the structure or material.

fatigue test
A test in which a specimen is subjected to measured reversals of stress or loading, i.e. alternate compression or tension repeatedly applied.

feathered propeller
A propeller whose blades have been rotated so that the leading and trailing edges are nearly parallel with the aircraft flight path to stop or minimise drag and engine rotation. Normally used to indicate shutdown of reciprocating or turboprop engine due to malfunction.

Federal Aviation Regulations
Regulations applicable to the operations of American registered aircraft, airspace and aviation personnel.

Federal Communications Commission
The controlling authority for radio communications in the United States.

Fédération Aeronautique Internationale
Formed to encourage growth of aeronautics and astronautics and promote all aviation sports throughout the world by championships and international competitions thereby underlining the essentially international spirit of aeronautics. Homologation of world aeronautical and astronautical records.

Federation of Air Transport User Representatives in the European Community
Formed to promote and encourage the establishment and maintenance in each Member State of a Committee whose purpose shall be to further the interests of all kinds of users of air transport.

feeder route [US]
A route depicted on instrument approach procedure charts to designate routes for aircraft to proceed from the en-route structure to the initial approach fix. See also INSTRUMENT APPROACH PROCEDURE.

fenestron (helicopter)
An Aerospatiale tail rotor, with many small blades shrouded in the centre of the tail fin, and seen most notably on Gazelle and Dauphin helicopters.

ferromagnetism
A high degree of magnetism found in iron, nickel, cobalt and various alloys.

ferry flight
A flight for the purpose of:

(1) Returning an aircraft to base.
(2) Delivering an aircraft from one location to another.
(3) Moving an aircraft to and from a maintenance base.

Ferry flights, under certain conditions, may be conducted under terms of a special flight permit.

ferry fuel tank
An additional tank (or tanks) carried for a ferry flight over a range greater than normal.

fibre optics
Optical equipment using thin glass fibres along which light is passed by means of internal reflection.

field
The region (volume) in which it is possible to measure the forces between different bodies which are not in contact (magnetic, gravitational, etc.)

field elevation   See AIRPORT ELEVATION.

field coil
Used for magnetising a core to create an electromagnet; used in generators, alternators and electric motors.

field length
A distance required during takeoff and landing in relation to accelerate/stop distance for a rejected takeoff and other operations when specified in the aircraft flight manual/pilot's operation manual.

filament
A very thin wire with a high melting point, usually tungsten (wolfram) at 3410°C, designed to supply incandescence for lights.

filed flight plan
The flight plan as filed with an ATS unit by the pilot or his designated representative, without any subsequent changes.
Note: When the word 'message' is used as a suffix to this term, it denotes the content and format of the filed flight plan data as transmitted.

fillet
An aerodynamic fairing.

filling
The occurrence of increasing pressure in the centre of a low pressure system. Filling is the opposite of deepening.

fin
A fixed vertical surface affecting the stability of an aircraft.

final
Commonly used to mean that an aircraft is on the final approach course or is aligned with a landing area. See also FINAL APPROACH COURSE, FINAL APPROACH IFR, TRAFFIC PATTERN and SEGMENTS OF AN INSTRUMENT APPROACH PROCEDURE.

final approach [ICAO]
That part of an instrument approach procedure from the time the aircraft has:

(1) Completed the last procedure turn or base turn, where one is specified; or
(2) crossed a specified fix; or
(3) intercepted the last track specified for the procedure, until it has crossed a point in the vicinity of an aerodrome from which:

  (i) a landing can be made; or
  (ii) a missed approach procedure is initiated.

Note: In UK legislation the point where the final 'approach to landing' begins is 1000 ft above Decision Height.

final approach course
A straight line extension of a localiser, a final approach radial/bearing or a runway centreline, all without regard to distance. See also FINAL APPROACH IFR and TRAFFIC PATTERN.

final approach F/A mode
The condition of DME/P operation which supports flight operations in the final approach and runway regions.

final approach fix
The designated fix from or over which the final approach (IFR) to an airport is executed. The final approach fix identifies the beginning of the final approach segment of the instrument approach. See also FINAL APPROACH POINT, GLIDESLOPE INTERCEPT ALTITUDE and SEGMENTS OF AN INSTRUMENT APPROACH PROCEDURE.

final approach IFR
The flight path of an aircraft that is inbound to an airport on a final instrument approach course, beginning at the final approach fix or point and extending to the airport or the point where a circle-to-land manoeuvre or a missed approach is executed. See also FINAL APPROACH COURSE, FINAL APPROACH FIX, FINAL APPROACH POINT and SEGMENTS OF AN INSTRUMENT APPROACH PROCEDURE.

final approach point
The point, within prescribed limits of an instrument approach procedure, where the aircraft is established on the final approach and a final approach descent may be

commenced. A final approach point is applicable only in non-precision approaches where a final approach fix has not been established. In such instances the point identifies the beginning of the final approach segment of the instrument approach. See also FINAL APPROACH FIX, GLIDESLOPE INTERCEPT ALTITUDE and SEGMENTS OF AN INSTRUMENT APPROACH PROCEDURE.

final approach segment
That segment of an instrument approach procedure in which alignment and descent for landing are accomplished. See also SEGMENTS OF AN INSTRUMENT APPROACH PROCEDURE.

final approach VFR   See TRAFFIC PATTERN.

final controller
The controller providing information and final approach guidance during PAR and ASR approaches utilising radar equipment. See also RADAR APPROACH.

fine pitch
The blade angle of a propeller which gives the best performance during a takeoff or go around.

fireproof
(1) With respect to materials and parts used to confine fire in a designed fire zone, the capacity to withstand, at least as well as steel in dimensions appropriate for the purpose for which they are used, the heat produced when there is a severe fire of extended duration in that zone.
(2) With respect to other materials and parts, the capacity to withstand the heat associated with fire at least as well as steel in dimensions appropriate for the purpose for which they are used.

fire resistant
(1) With respect to sheet or structural members, the capacity to withstand the heat associated with fire at least as well as aluminium alloy in dimensions appropriate for the purpose for which they are used.
(2) With respect to fluid-carrying lines, fluid system parts, wiring, air ducts, fittings and powerplant controls, the capacity to perform the intended functions under the heat and other conditions likely to occur when there is a fire at the place concerned.

fires
Fires are classified according to the type(s) of material being burned. Class A fires consist of wood, cloth, paper, etc. Extinguishing agents used for class A fires:
(1) 'Soda acid type', producing water and carbon dioxide under pressure by mixing sulphuric acid with sodium hydrogen carbonate.
(2) 'Hologenated hydrocarbon type', BCF bromochlorodifluormethane, CB chlorobromomethane.
Class B fires consist of petrol, paraffin, diesel, paints, etc. Extinguishing agents used for class B fires:

(1) BCF and CB.
(2) Carbon dixoide.
(3) Air foam using protein.

(4) Chemical foam using sodium hydrogen carbonate mixed with aluminium sulphate.

firewall
Fire resistant bulkhead between the engine and the rest of the aircraft.

firing order
The sequence in which the cylinders in an engine fire, i.e. 1–3–4–2 or 1–5–3–6–2–4.

first gust
The leading edge of the spreading downdraft, from an approaching thunderstorm.

first officer
The pilot who is second in command. Also may be called co-pilot.

fix
A definite position of an aircraft, determined visually; by the intersection of bearings or radials; by other lines of position; by VOR radial and DME distance; or by other means, without reference to any former position.

fixed landing gear
Landing gear which does not retract.

fixed-wing special IFR [US]
Aircraft operating in accordance with a waiver and a letter of agreement within control zones specified in FARs. These operations are conducted by IFR qualified pilots in IFR equipped aircraft and by pilots of agricultural and industrial aircraft.

fix, outer [US]
A fix in the destination terminal area other than the approach fix, to which aircraft are normally cleared by an ARTCC, and from which aircraft are cleared to the approach fix or final approach course.

fix, running
A fix obtained by moving forward the first of two lines of position; the distance and direction made good between the taking of two observations, to intersect with the second line of position.

flag
A warning device incorporated in certain airborne navigation and flight instruments indicating that:
(1) instruments are inoperative or not operating satisfactorily, or
(2) signal strength or quality of the received signal falls below acceptable values.

flagalarm   See FLAG.

flameout
Unintended loss of combustion in turbine engines resulting in the loss of engine power.

flame resistant
Not susceptible to combustion to the point of propagating a flame, beyond safe limts, after the ignition source is removed.

flammable
With respect to a fluid or gas, susceptible to igniting readily or to exploding.

flap
An auxiliary wing surface basically designed to increase lift and drag. Many types of flap design are in use and each has its own particular advantages.

flap extended speed
The highest speed permissible with wing flaps in a prescribed extended position.

flapping hinge
Helicopter rotor blades fly both high and low during their rotation. To allow them to do this freely, the blades can be mounted on a horizontal pin at the blade root. This pin is known as the flapping hinge and the action of the blade moving up and down on this pin is called flapping.

flare
The final phase of a landing during which the rate of descent is reduced with height.

flare path
Line of lights down one or both sides of a runway.

flares
High intensity magnesium lights which may be dropped by parachutes for emergency night landings.

flash over
Discharge through air between conductors across which a large potential exists.

flash point
The lowest temperature at which a fuel produces a vapour which can be ignited by a spark or small flame.

flash resistant
Not susceptible to burning violently when ignited.

flat spin
A spin at a large mean angle of attack, with the aircraft in an almost horizontal position. Recovery from this manoevre can be difficult, as the aircraft is in a fully stalled condition and the control surfaces are largely ineffective.

flat zone
A zone within an indicated course sector or an indicated ILS glide path sector in which the slope of the sector characteristic curve is zero.

flexible blade (helicopter)
A system where the rotor blades have neither pivots nor hinges. The rotor blades themselves are flexible and each blade has a controllable tab-type airfoil attached to its trailing edge. Movement of these tabs twists the rotor blades and controls their angle of pitch.

flight
In the case of a piloted flying machine, from the moment when, after the embarkation of its crew for the purpose of taking off, it first moves under its own power until the moment when it next comes to rest after landing.

flight check   See FLIGHT INSPECTION.

flight check [US]
A call-sign prefix used by FAA aircraft engaged in flight inspection/certification of navigational aids and flight procedures. The word 'recorded' may be added as a suffix. e.g. 'Flight check 320 recorded', to indicate that an automated flight inspection is in progress in terminal areas. See also FLIGHT INSPECTION.

flight crew member
A licensed crew member charged with duties essential to the operation of an aircraft during flight time.

flight data
Data regarding the actual or intended movement of aircraft, normally presented in coded or abbreviated form.

flight data recorder
A recorder system fitted in public transport aircraft providing full information on the aircraft operation, or about conditions encountered.

flight deck
The cockpit or compartment used by the flight crew on a large aircraft.

flight director system   See INTEGRATED FLIGHT SYSTEM.

flight duty period
The total time from the moment a flight crew member commences duty, immediately subsequent to a rest period and prior to making a flight or a series of flights, to the moment he is relieved of all duties, having completed such flight or series of flights.

flight information
Information useful for the safe and efficient conduct of flight, including information on air traffic, meteorological conditions, aerodrome conditions or air route facilities.

flight information centre
A unit established to provide flight information service and alerting service.

flight information region
An airspace of defined dimensions within which flight information service and alerting services are provided.

(1) Flight information service. Gives advice and information useful for the safe and efficient conduct of flights.
(2) Alerting service. Notifies appropriate organisation regarding aircraft in need of search and rescue aid and assists such organisations as required.

flight information service
A service provided for the purpose of giving advice and information useful for the safe and efficient conduct of flights.

flight inspection
(1) In-flight investigation and evaluation of a navigational aid to determine whether it meets established tolerances. See also NAVIGATIONAL AID.
(2) Periodic check on flight crew and air traffic controllers, as required by aviation administrators to determine maintenance of competency.

flight level
A level of constant atmospheric pressure related to a reference datum of 29.92 inches of mercury. Each is stated in three digits that represent hundreds of feet. For example, flight level 250 represents a barometric altimeter indication of 25 000 ft; flight level 255, an indication of 25 500 ft.

flight levels [ICAO]
Surfaces of constant atmospheric pressure that are related to a specific pressure datum of 1013.2 mb (29.92 in) and are separated by specific pressure intervals.

flight line
A term used to describe the precise movement of a civil photogrammetric aircraft along a predetermined course(s), at predetermined altitude, during the actual photographic run.

flight log
A device which records an aircraft's flight path in the horizontal plane, for example a roller map.

flight panel
A panel of aircraft flight instruments giving sufficient indications to enable the pilot to fly without external references.

flight path
A line, course or track along which an aircraft is flying or is intended to be flown. See also COURSE and TRACK.

flight plan
Specified information relating to the intended flight of an aircraft, filed orally or in writing with an ATC facility.

flight plan data
Data selected from the flight plan for purposes of processing, display or transfer.

flight progress board
A board designed and used for the tabular display of flight data. Used by air traffic control staff.

flight progress display
A display of data from which the actual and intended progress of flights may be readily determined.

flight progress strip
Strip used for the display of flight data on a flight progress board.

flight recorder
A general term applied to any instrument or device that records information about the performance of an aircraft in flight, or about conditions encountered in flight. Flight recorders may make records of airspeed, outside air temperature, vertical acceleration, engine RPM, manifold pressure and other pertinent variables for a given flight.

flight recorder [ICAO]
Any type of recorder installed in the aircraft for the purpose of complementing accident/incident investigations.

flight release certificate [UK]
A certifying statement issued by a licensed aircraft maintenance engineer, or other approved person, declaring an aircraft with a Permit to Fly as fit for flight.

Flight Safety Committee [UK]
Objectives: founded in 1959, to collect and disseminate flight safety information and to foster interest in and work for the advancement of flight safety in all sections of aviation. The Committee is an entirely independent body supported by a number of aviation organisations.

flight service station [US]
Air traffic facilities that provide pilot briefing, en-route communications and VFR search and rescue services, assist lost aircraft and aircraft in emergency situations, relay ATC clearances, originate Notices to Airmen, broadcast aviation weather and NAS information, receive and process IFR flight plans and monitor navaids. In addition, at selected locations, FSSs provide en-route flight advisory service (flight watch), take weather observations, issue airport advisories and advise Customs and Immigration of transborder flights.

flight simulator   See SYNTHETIC FLIGHT TRAINER.

flight standards district office [US]
An FAA field office serving an assigned geographical area staffed with flight standards personnel who serve the aviation industry and the general public on matters relating to the certification and operation of air carrier and general aviation aircraft. Activities include general surveillance of operational safety, certification of airmen and aircraft, accident prevention, investigation, etc.

flight status
An indication of whether a given aircraft requires special handling by air traffic service units or not.

flight test
A flight for the purpose of:
(1) Investigating the operation/flight characteristics of an aircraft or aircraft component.
(2) Evaluating an applicant for a pilot certificate, licence or rating.

flight time
The total time from the moment an aircraft first moves under its own power for the purpose of taking off until the moment it comes to rest at the end of the flight. *Note:* Flight time as here defined is synonymous with the term 'block to block' time or 'chock to chock' time in general usage which is measured from the time the aircraft moves from the loading point until it stops at the unloading point.

flight time limitation board
A body set up in the United Kingdom with the object of recommending regulations concerning the hours worked and flight times of aircrew employed in commercial air transport operations.

flight visibility
The average forward horizontal distance, from the cockpit of an aircraft in flight, at which prominent unlighted objects may be seen and identified by day and prominent lighted objects may be seen and identified by night.

flight watch [US]
A shortened term for use in air–ground contacts on frequency 122.0 MHz to identify the flight service station providing en-route flight advisory service, e.g. 'Oakland Flight Watch'. See also EN-ROUTE FLIGHT ADVISORY SERVICE.

float plane
A marine aircraft with one, two or sometimes more floats and a conventional fuselage.

flow control
Measures designed to adjust the flow of traffic into a given airspace, along a given route or bound for a given airport, so as to ensure the most effective utilisation of the airspace. See also QUOTA FLOW CONTROL.

flow line
A streamline.

fluid ounce
Measure of volume of liquids = 28.41 cc.

fluorescence
Absorption of light of one wavelength and emission of light of another wavelength.

fluorescent lamp
A glass tube with an inside coating of a fluorescent substance which converts ultraviolet light into visible light. The ultraviolet light is produced by the action of electrons on mercury vapour.

flush antenna
An antenna which conforms to the shape of the aircraft.

flutter
A rapid, unstable oscillation due to alternating forces.

flux
The rate of flow of energy (with or without mass) per unit area.

flux density
For magnetic flux the unit of density is the tesla (T) = $Wb\,m^{-2}$.

fly by wire
A flight control system using computers and electrical signalling for the operation of mechanical control surfaces.

fly heading (degrees)
Informs the pilot of the heading he should fly. The pilot may have to turn to, or continue on, a specfic compass direction in order to comply with the instructions. The pilot is expected to turn in the shorter direction to the heading, unless otherwise instructed by ATC.

flying duty period
Any time during which a person operates in an aircraft as a member of its crew. It starts when the crew member is required by an operator to report for a flight, and finishes at on-chocks or engines off, or rotors stopped, on the final sector.

focal length
The light distance between the lens and film of a camera.

foehn
A warm, dry, downslope wind, the warmth and dryness being due to adiabatic compression upon descent. Characteristic of mountainous regions. See also ADIABATIC PROCESS, CHINOOK, SANTA ANA.

fog
A cloud at or near the earth's surface. Fog consists of numerous droplets of water, which individually are so small that they cannot readily be distinguished by the naked eye.

foot
Unit of length = 0.3048 m.

foot-pound
Unit of energy or work. Work done by a force of one pound over a distance of one foot.

force
Mass × Acceleration measured in newtons. N = $kg\,ms^{-2}$.

foreign air carrier [US]
Any person other than a citizen of the United States, who undertakes directly, by lease or other arrangement, to engage in air transportation.

foreign air commerce [US]
The carriage by aircraft of persons or property for compensation or hire, or the carriage of mail by aircraft, or the operation or navigation of aircraft in the conduct or furtherance of a business or vocation in commerce between a place in the United States and place outside thereof, whether such commerce moves wholly by aircraft or partly by aircraft and partly by other forms of transportation.

foreign air transportation [US]
The carriage by aircraft of persons or property as a common carrier for compensation or hire, or the carriage of mail by aircraft, in commerce, between a place in the United States and any place outside of the United States, whether that commerce moves wholly by aircraft or partly by aircraft and partly by other forms of transportation.

foreign airworthiness directive
A directive covering modifications and inspections which have been classified as mandatory by foreign airworthiness authorities.

formation flight
More than one aircraft that, by prior arrangement between the pilots, operate as a single aircraft with regard to navigation and position reporting. Separation between aircraft within the formation is the responsibility of the flight leader and the pilots of other aircraft in the flight. This includes transition periods when aircraft within the formation are manoeuvring to attain separation from each other to effect individual control and during join-up breakaway.

Fortin barometer
A very accurate mercury barometer.

fossil fuels
Coal, oil, natural gas, all from decomposed organisms rich in carbon and hydrogen.

Foucault pendulum
A heavy weight suspended from a very long wire which will swing slowly for several hours. The apparent change of direction of its swing is evidence of the rotation of the earth beneath it.

FPS system
The foot/pound/second system of measurement still used in parts of the United States.

fractus
Clouds in the form of irregular shreds, appearing as if torn and with a clearly ragged appearance. Applies only to stratus and cumulus, i.e. cumulus fractus and stratus fractus.

frangibility
A characteristic of an object to retain its structural integrity and stiffness up to a desired maximum load, but, on impact from a greater load, to break, distort or yield in such a manner as to present the minimum hazard to aircraft.

free airport
An international airport at which, provided they remain within a designated area until removal by air to a point outside the territory of the State, crew, passengers, baggage, cargo, mail and stores may be disembarked or unladen, may remain and may be trans-shipped, without being subjected to any customs charges or duties and, except in special circumstances, to any examination.

free balloon
A balloon which, when in flight, is not attached to any form of retaining device to the surface.

freedoms
There are five basic freedoms relating to aircraft international movements. These are negotiated by governments of countries.

(1) First freedom. The right to fly across the territory of another country under the terms of the agreement.
(2) Second freedom. The right to make technical (non-traffic) stops in that country.
(3) Third freedom. The right to disembark passengers or cargo in that country and embarked in the home country of the aircraft operator.
(4) Fourth freedom. The right to pick up passengers or cargo in that country and destined for the airline operator's home country.
(5) Fifth freedom. The right to pick up passengers or cargo in any country for transportation to any third country.

free electron
An electron that is not bound to any atomic structure and is free to move in an electric field.

free flight
A flight without guidance.

free wheeling unit (helicopter)
A free wheeling unit or sprag clutch is a device which automatically disengages the engine from the rotor system in the event of an engine malfunction or sudden retarding of the throttle. Thus the blades will continue to rotate without drag from the engine. When proper control procedures are employed, the free wheeling unit enables the pilot to perform a safe, autorotational landing.

freezing level
A level in the atmosphere at which the temperature is 0°C (32°F).

freezing point
The temperature at which a substance may be either liquid or solid (at a pressure of one atmosphere, i.e. 760 mm mercury).

free zone
An area where merchandise, whether of domestic or foreign origin, may be admitted, deposited, stored, packed, exhibited, sold, processed or manufactured, and from which such merchandise may be removed to a point outside the territory of the State without being subjected to customs duties or internal consumer taxes or, except in special circumstances, to inspection. Merchandise of domestic origin admitted into a free zone may be deemed to be exported.

frequency
Cycles per second of wave motion, i.e. velocity/wavelength, measured in Hertz (Hz).

frequency band
Internationally defined bands of radio frequencies between agreed limits.

frequency channel
A continuous portion of the frequency spectrum appropriate for a transmission utilising a specified class of emission.
Note: The classification of emissions and information relevant to the portion of the frequency spectrum appropriate for a given type of transmission (bandwidths) is specified in the ITU Radio Regulations, Article 4, RR 264 to RR 273 inclusive.

frequency modulation
The modulation of the frequency of the carrier wave instead of its amplitude.

frequency pairing
The permanent association of frequencies in different systems such as VOR/DME and Localiser/Glideslope.

frequency spectrum
The distribution of signal amplitudes as a function of frequency.

Fresnel
A unit of frequency = 1 million megahertz, MHz $\times 10^6$ (Hz $\times 10^{12}$).

Fresnel lens
Used in navigation and headlights, a lens whose surface is a composite of many smaller lenses giving a short focal length.

friction horsepower
The indicated horsepower minus brake horsepower.

frise type aileron
An aileron designed and pivoted so that when it is raised its nose projects below the main plane. The nose does not project above the main plane when the aileron is lowered.

front
The zone of transition between two air masses of different density. It is the surface of air dividing the two air masses, extending from the ground to the 'top' of the air masses; that is, up to some level beyond which it is impossible to distinguish one air mass from the other. This is called the frontal surface, and the line where it meets the ground is called a front. The surface is not vertical, but slopes gently upwards from the cold to the warm side, so that the cold air forms a very flat wedge underneath the warm. This formation arises from the difference of density between the air masses, the lighter warm air tending to float on top of the denser cold air.

front course sector
The course sector which is situated on the same side of the localiser as the runway.

frontogenesis
The beginning or creation of a front.

frontolysis
The destruction and dying of a front.

frost
Crystals of ice formed like dew, but at a temperature below freezing.

fruit
Non-synchronous SSR replies.

fuel advisory departure [US]
Procedures to minimise engine running time for aircraft destined for an airport experiencing prolonged arrival delays.

fuel dumping
Airborne release of unusable fuel. This does not include the dropping of fuel tanks. See JETTISONING OF EXTERNAL STORES.

fuel exhaustion
When all fuel carried aboard the aircraft has been used up.

fuel injection system
Direct fuel injection systems meter fuel directly into the cylinder and have many advantages over a conventional carburettor system. They reduce the possibility of induction icing since a drop in temperature due to fuel vaporisation (which causes approximately 70% of the temperature drop during mixing of air and fuel at cruise power setting) occurs in or near the cylinder. In addition to this, a direct fuel injection system provides better engine acceleration and improved fuel distribution, leading to greater economy of operation. Fuel injection systems vary in detail but basically they consist of a fuel injector, flow divider and fuel discharge nozzle. The fuel injector monitors the volume of air entering the system and meters the fuel flow. The flow divider keeps the fuel under pressure and distributes the fuel to the various cylinders at the rate required for different engine speeds. It also shuts off the individual nozzle lines when the idle cut-off control is used.

fuel siphoning
Unintentional release of fuel caused by overflow, puncture, loose fuel cap, etc.

fuel starvation
Fuel is available but for some reason is not reaching the engine(s).

fuel venting   See FUEL SIPHONING.

full-wave rectifier
Changes the negative halves of alternating current waves into positive, thus giving a type of direct current.

fully articulated blade (helicopter)
A helicopter rotor system of three or more blades that are individually hinged so that each blade has freedom to move both up and down (flapping), forwards and backwards (lead-lag) and to change pitch.

fully-automatic relay installation
A teletypewriter installation where interpretation of the relaying responsibility in respect of an incoming message and the resultant setting-up of the connections required to effect the appropriate retransmissions is carried out automatically, as well as all other normal operations of relay, thus obviating the need for operator intervention, except for supervisory purposes.

funnel cloud
A tornado cloud or vortex cloud extending downward from the parent cloud but not reaching the ground.

fuse
Section of a circuit made of wire with a low melting point which melts and breaks the circuit when the current is above limits.

fuselage
The main structural body of an aircraft to which the mainplanes, tail unit, etc. are attached.

G

gain
A general term used to denote an increase in signal power.

galaxy
Clusters of stars, about 10^{11} stars in each; some spiral shaped, some elliptical and some irregularly shaped. The sun is in the galaxy called the 'milky way' which is a disc-shaped spiral about 10^5 light years in diameter. The sun is about three fifths of the way from the centre. The nearest galaxy to the milky way is 16×10^5 light years away.

gallon
Unit of volume. A British (or imperial) gallon = volume of 10 lb water, or 4.54609 litres. A US gallon is smaller, 0.83269 of an imperial gallon, or 3.79 litres.

galvanised iron
Sheet iron (often corrugated) coated with zinc to prevent corrosion.

galvanometer
Simple magnetic instrument for detecting very small electric currents.

garbling
The degradation of code information due to the simultaneous presence in a decoder of overlapping reply pulse trains, applicable to SSR interrogations.

gascolator
In light aircraft, a main fuel strainer fitted at the lowest point of the fuel system. Fuel which has left the tank flows through this strainer before reaching the carburettor; its purpose is to trap water and other foreign matter.

gas oil
Diesel oil

gasoline
US term for petrol.

gas turbine
In turbopropeller and turbojet engines, the air is compressed before the fuel is injected and ignited.

gate
A point of access to the apron from the terminal at an airport.

gate hold procedures
Procedures at selected airports to hold aircraft at the gate or other ground location whenever departure delays exceed or are anticipated to exceed five minutes. The

sequence for departure will be maintained in accordance with initial call-up unless modified by flow control restrictions. Pilots should monitor the ground control/clearance delivery frequency for engine startup advisories or new proposed start time if the delay changes. See also FLOW CONTROL.

Gay-Lussac's law
States that when gases combine they do so in a simple ratio by volume.

Gegenschein counter glow
The reflection of the sun's rays by meteoric particles in space, seen at night, sometimes, opposite the sun.

Geiger counter
An instrument which detects alpha, beta and gamma rays.

gelignite
An explosive mixture of potassium nitrate, cellulose nitrate and nitroglycerin on a base of wood pulp.

general aviation [ICAO]
Prior to 1990 the definition of general aviation activities was 'all civil aviation operations other than scheduled air services and non-scheduled air transport operations for remuneration or hire.' This definition has subsequently been changed to: An aircraft operation other than a commercial air transport operation or an aerial work operation.

General Aviation District Office [US]
An FAA field office serving a designated geographical area, staffed with flight standards personnel, who have responsibility for serving the aviation industry and the general public on all matters relating to the certification and operation of general aviation aircraft.

General Aviation Manufacturers and Traders Association [UK]
Formed to represent the interests of the UK aviation manufacturing companies and sales organisations and other supporting services and general aviation commercial operations. The main objectives are to ensure the healthy growth of general aviation throughout the United Kingdom, to protect the interests of the industry and to promote the use of general aviation aircraft.

General Aviation Safety Committee [UK]
A non-profit making independent organisation consisting of representatives of all elements of UK general aviation with the aim of fostering the development of general aviation in the United Kingdom along safe lines by encouraging competence among UK general aviation pilots and operators.

generator (dynamo)
A machine which converts mechanical energy into electrical energy by forcing conductors across magnetic lines of force.

geocentric
Measured from the centre of the earth.

geodesic line
The shortest distance between two points on the surface of a sphere.

geomagnetism
Terrestrial magnetism; the earth's magnetic field.

geometric dilution of precision   See DILUTION OF PRECISION.

geostar [US]
A multi-purpose satellite system using fixed-orbit satellites.

geostationary orbit
The orbit in which a satellite remains over the same point on the earth's surface.

geostrophic force
Air in motion behaves as if it were acted upon by a force at right angles to its direction of motion, and acts from left to right in the northern hemisphere and from right to left in the southern hemisphere. This force is known as the geostrophic force, and it can be shown mathematically that its value is proportional to the wind speed and the sine of the latitude.

geosynchronous
Revolving at the same angular speed as the earth.

giant star
A star 10 to 100 times larger than the sun.

giga
Prefix for 10^9.

gimbal
A frame in which a gyro is mounted so as to allow freedom of movement about an axis perpendicular to the gyro spin axis.

gimbal freedom
The maximum angular displacement about the gyroscope axis before toppling occurs. Dependent upon the construction of the gyroscope system.

glass cockpit
A cockpit in which electronic visual displays are used instead of conventional instruments.

glaze
A coating of ice, generally clear and smooth, formed by freezing of supercooled water on a surface. See also CLEAR ICING.

glidepath [ICAO]
A descent profile determined for vertical guidance during a final approach.

glidepath (on/above/below)
Used by ATC to inform an aircraft making a PAR approach of its vertical position (elevation) relative to the descent profile. The terms 'slightly' and 'well' are used to describe the degree of deviation, e.g. 'slightly above glidepath'. Trend information

is also issued with respect to the elevation of the aircraft and may be modified by the terms 'rapidly' and 'slowly', e.g. 'well above glidepath, coming down rapidly'.

glider
A heavier-than-air aircraft that is supported in flight by the dynamic reaction of the air against its lifting surfaces and whose free flight does not depend upon an engine.

glider flight time
The total time occupied in flight, whether being towed or not, from the moment the glider first moves for the purpose of taking off until the moment it comes to rest at the end of the flight.

glide ratio
The ratio of horizontal distance covered to the amount of height lost.

glideslope
Provides vertical guidance for aircraft during approach and landing. The glideslope consists of the following:
(1) Electronic components emitting signals that provide vertical guidance by reference to airborne instruments during instrument approaches such as ILS, or MLS; or
(2) Visual ground aids such as VASI that provide vertical guidance for VFR approach or for the visual portion of an instrument approach and landing.

glideslope intercept altitude
The minimum altitude of the intermediate approach segment prescribed for a precision approach that assures required obstacle clearance. It is depicted on instrument approach procedure charts. See INSTRUMENT LANDING SYSTEM and SEGMENTS OF AN INSTRUMENT APPROACH PROCEDURE.

global positioning system
A world-wide system by which the user can derive position by receiving signals from navstar satellites.

glycol
Used as anti-freeze; colourless, viscous, boiling point 197°C.

gnomonic
A chart projection on which all great circles are exactly represented by straight lines, and all straight lines are great cirles.

go ahead
Proceed with your message. Not to be used for any other purpose.

go around
Instructions for a pilot to abandon his approach to landing. Additional instructions may follow. Unless otherwise advised by ATC, a VFR aircraft or an aircraft conducting a visual approach should overfly the runway while climbing to traffic pattern altitude and enter the traffic pattern via the crosswind leg. A pilot on an IFR flight plan making an instrument approach should execute the published missed

approach procedure or proceed as instructed by ATC, e.g. 'Go around' (additional instructions, if required). See also LOW APPROACH and MISSED APPROACH.

goosenecks
Paraffin flares used as runway lights.

government aerodrome
Any aerodrome in the United Kingdom which is in the occupation of any government department or visiting force.

governor
A device for regulating the speed of a machine.

gradient
The ratio of vertical distance to horizontal distance (whereas 'slope' is the ratio of vertical distance to the actual path: the hypoteneuse). In meteorology, a horizontal decrease in value per unit distance of a parameter in the direction of maximum decrease; most commonly used with pressure, temperature and moisture.

gradient tints
A system of representing chart relief or changes in elevation by a series of colour tints ranging from green at sea level to dark brown at the higher elevations.

gradient wind
The circular path of the air due to geostrophic force causes a small centripetal effect upon the air particles, known as the cyclostrophic component. It affects wind speed. When this is taken into account the result is known as the gradient wind. It becomes significant in relation to tropical storms, tornados, etc.

gradual
In meteorology forecasts refers to gradual change at approximately constant rate throughout the period, or during a specified part thereof.

graduations
The markings (divisions and subdivisions) of the scale of an instrument.

Graham's law
The rate of diffusion of a gas is inversely proportional to the square root of its density.

gramme (gram)
A unit of mass equal to one thousandth of a kilogramme. Gram weight or gram force is the force resulting from gravity $\times$ 1 gramme = 981 dynes or 0.0353 oz. 1 lb = 453.6 g.

granular snow
Frozen drizzle, a form of precipitation which may be associated with stratus type cloud. It consists of small, opaque grains.

graticule
A network of lines, e.g. longitude and latitude, shown on aeronautical navigation charts.

gravity
The attraction between two masses:

$$F = \frac{G.m_1 . m_2}{d^2}$$

Where F = force, m, and m_2 are the two masses. d is the distance between them and G is the gravitational constant $6.664 \times 10^{-11} \, Nm^2 kg^{-2}$

The earth's gravity causes an acceleration of $9.81 \, m.s^{-2}$ The moon's gravity is about one sixth of the earth's gravity on its surface.

gray
The energy imparted by ionising radiation to a mass of matter corresponding to 1 joule per kilogram.

grease
A lubricant made from hydrocarbon soaps emulsified in petroleum oils.

great circle
The circle that would be formed by bisecting the globe with a plane passing through the centre.

greenhouse effect
The sun's short-wave radiation passes through the earth's atmosphere but the re-radiation from the earth's surface is long-wave and this is absorbed by the CO_2 in the atmosphere, thus stopping the energy.

Greenwich
The location of the principal British observatory, near London. In most countries, longitude is reckoned east or west from the meridian passing through this observatory. Hence it is the earth's prime meridian (0) and the origination of Greenwich Mean Time.

grid
A network of horizontal and vertical lines superimposed on a map (sometimes called a 'graticule'); processed meteorological data in the form of grid point values (in aeronautical meteorological code).

grid point data in digital form
Computer processed meteorological data for a set of regularly spaced points on a chart, for transmission from a meteorological computer to another computer in a code form suitable for automated use.
Note: In most cases such data is transmitted on medium or high speed telecommunications channels.

grid point data in numerical form
Processed meteorological data for a set of regularly spaced points on a chart, in a code form suitable for manual use.
Note: In most cases such data is transmitted on low speed telecommunications channels.

ground clutter
A pattern produced on the radar scope by ground returns that may degrade other radar returns in the affected area. The effect of ground clutter is minimised by the use of moving target indicator (MTI) circuits in the radar equipment resulting in a radar presentation that displays only targets that are in motion. See also CLUTTER.

ground controlled approach
A radar approach system operated from the ground by air traffic control personnel transmitting instructions to the pilot by radio. The approach may be conducted with surveillance radar (ASR) only, or with both surveillance and precision approach radar (PAR). Usage of the term 'GCA' by pilots is discouraged except when referring to a GCA facility. Pilots should specifically request a 'PAR' approach when a precision radar approach is desired or request an 'ASR' or 'surveillance' approach when a non-precision radar approach is desired. See also RADAR APPROACH.

ground effect (aeroplane)
(1) When an aeroplane in flight gets within several feet from the ground surface, a change occurs in the three dimensional flow pattern around the aeroplane because the vertical component of the airflow around the wing is restricted by the ground surface. This alters the wing's upwash, downwash and wingtip vortices. These general effects due to the presence of the ground are referred to as ground effect. Ground effect, then, is due to the interference of the ground (or water) surface with the airflow patterns about the aeroplane in flight. The reduction in induced flow due to ground effect causes a significant reduction in induced drag, but causes no direct effect on parasite drag. As a result of the reduction in induced drag, the thrust required at low speeds will be reduced.
(2) Effects caused by interference from the surface of the earth on radar screens, radio navaids and other EM systems.

ground effect (helicopter)
This results from the downwash of the blades and produces a circular volume of semi-confined air between the ground and the rotor blades, providing additional lift. The effect is of practical value up to a height equal to half the diameter of the rotor system, although its influence can be felt at even greater heights than this. Ground effect can be partially dissipated by uneven ground, dense ground cover and moderate to high wind conditions.

ground equipment
Articles of a specialised nature for use in the maintenance, repair and servicing of an aircraft on the ground. Includes testing equipment, and cargo- and passenger-handling equipment.

ground fog
In the United States, a fog that conceals less than 0.6 of the sky and is not contiguous with the base of clouds.

ground loop
An uncontrolled turn while taking off or during the landing roll. More commonly occurring with tailwheel-type aeroplanes.

ground movement control
A common term for the air traffic control position which controls aircraft taxi-ing and ground vehicle movements.

ground movement controller
A member of the ATC staff who is responsible for the safe and orderly flow of aircraft movements on the ground.

ground plot   See AIR PLOT.

ground proximity warning
A system which warns pilots that they are dangerously close to the surface.

ground resonance (helicopter)
Ground resonance can develop when a series of ground shocks causes the rotor head to become unbalanced. This is usually caused by an uneven or heavy landing in certain types of helicopters. The condition, if allowed to progress, can result in structural damage and even total disintegration of the helicopter in a very short space of time. The condition can be recognised by progressively violent oscillations. There are two accepted methods of recovery:

(1) With flight rpm still available, depart the ground by lifting into a hover.
(2) With low rpm, reduce collective pitch and close the throttle.

ground signals
Signals displayed on the surface of an aerodrome, usually situated near the control tower.

ground speed
The speed of an aircraft with reference to the surface of the earth.

ground-to-air communication
One-way communication from stations or locations on the surface of the earth to aircraft.

ground visibility
The prevailing horizontal visibility near the earth's surface.

ground waves
Radio waves that travel directly from transmitter to receiver without being reflected from the ionosphere.

Guild of Air Pilots and Air Navigators of London
Founded 1929; constituted a Livery Company of the City of London 10 April 1956. The main object is the achievement of air safety through the promotion of the highest standards of airmanship among pilots and navigators. Promotes the sound education and training of pilots and navigators by the award of scholarships and awards for achievement and issues Master Air Pilot and Master Air Navigator certificates. Through the Guild of Air Pilots Benevolent Fund it assists the dependants of pilots and navigators with education and in their old age.

Guild of Air Traffic Control Officers [UK]
Formed to promote honourable practice and maintain in the profession of air traffic control a high standard of efficiency and integrity dedicated to the safety of those who seek their livelihood or pleasure in the air.

Guild of Business Travel Agents [UK]
Formed in 1967, the Guild sets about improving standards, services and facilities among airlines, car hire companies, transportation companies, hotel groups and other principals on behalf of the business traveller. Nearly 50 members operate through over 1400 retail offices.

gust
A sudden brief increase in wind; gusts are usually reported when the variation in wind speed between peaks and lulls is at least 10 knots.

gust loading
The increased aerodynamic loads on an aircraft as a result of gusts.

gyrodyne
A rotorcraft whose rotors are normally engine-driven for takeoff, hovering and landing and for forward flight through part of its speed range, and whose means of propulsion, consisting usually of conventional propellers, is independent of the rotor system.

gyro horizon
A gyro-stabilised instrument providing a horizontal reference for aircraft in instrument flight, when the natural horizon is not visible.

gyroplane
A rotorcraft whose rotors are not engine-driven except for initial starting, but which are made to rotate by action of the air when the rotorcraft is moving; and whose means of propulsion, consisting usually of conventional propellers, is independent of the rotor system.

gyroscope
A rapidly spinning wheel mounted on gimbals. The axis and plane of rotation remain rigid in space (inertia) unless a force is applied to its axle. When a force is applied it causes 'precession', i.e. the resultant is as if the force were applied 90° forward in the direction of rotation.

gyroscopic precession
A characteristic of all rotating bodies, such as a gyroscope or helicopter's rotor system. Simply stated, when a force is applied to a spinning body, the resultant action will take place 90° later in the plane of rotation.

gyrosyn compass
An instrument combining a flux-gate compass with a directional gyro. The gyro is continuously and automatically reset by the flux-gate, resulting in steady and accurate compass readings without turning errors.

H

hail
Precipitation consisting of balls or irregular lumps of ice, often of considerable size. A single unit of hail is called a hailstone. Large hailstones usually have a centre surrounded by alternating layers of clear and cloudy ice. Hail falls almost exclusively in connection with thunderstorms.

half ILS glide path sector
The sector in the vertical plane containing the ILS glide path and limited by the loci of points nearest to the glide path at which the DDM is 0.0875.

halo
A prismatically coloured or whitish circle, or arcs of a circle, with the sun or moon at its centre; coloration, if not white, is from red inside to blue outside (opposite to that of a corona); fixed in size with an angular diameter of 22° (common) or 46° (rare); characteristic of clouds composed of ice crystals; valuable in differentiating between cirriform and forms of lower clouds.

handoff
An action taken to transfer the radar identification of an aircraft from one controller to another, if the aircraft is entering the receiving controller's airspace, and radio communications with the aircraft are being transferred.

hard copy
Readable pages, i.e. printed copy.

hard data
Data which remains in memory with power removed.

hardware
The sum total of components of a system which have a physical existence.

harmonics
Secondary frequencies occurring together with the basic wave frequency, i.e. a second harmonic has twice the frequency, a third harmonic has three times the frequency, etc.

have numbers
Used by pilots to inform ATC that they have received runway, wind and altimeter information only.

hazard beacon
An aeronautical beacon used to designate a danger to air navigation.

haze
A type of lithometeor composed of fine dust or salt particles dispersed through a portion of the atmosphere; particles are so small they cannot be felt or individually seen with the naked eye (as compared with the larger particles of dust), but diminish the visibility; distinguished from fog by its bluish or yellowish tinge.

heading
The direction in which the longitudinal axis of an aircraft is pointed, usually expressed in degrees from north (true, magnetic, compass or grid).

head up display
Equipment which allows information to be visually presented to the pilot while looking through the windscreen.

heat
The kinetic energy of atoms and molecules; moving from one place to another, rotating and vibrating and transmitted by conduction or convection.

(1) Radiant heat (Infra red radiation). Electromagnetic radiation with wave lengths between visible light and radio waves and not dependent upon the presence of matter.
(2) Specific heat. The heat lost or gained when the temperature of a substance changes without change of state.
(3) Latent heat. The heat lost or gained when the state of a substance changes without change of temperature.

heat exchanger
A system designed to transfer heat from one fluid to another without the two fluids coming into contact.

heat shield
The heat insulating surface of a spacecraft.

heavier-than-air aircraft
Any aircraft deriving its lift in flight chiefly from aerodynamic forces.

Heaviside-Kennelly layers
That part of the ionosphere (from 90 to 150 km amsl) that reflects the 'sky waves' of radio transmissions.

heavy aircraft [ICAO]
For the purposes of wake turbulence categorisation, aircraft with a maximum takeoff weight of 136 000 kg or greater are categorised as 'heavy'.

hectopascal
The approved ICAO unit for the measuring and reporting of atmospheric pressure. The value of the hectopascal is identical to the millibar.

height
The vertical distance of a level, a point or an object considered as a point, measured from a specified datum.

height above airport
The minimum descent altitude (MDA) above the published airport elevation; for use in conjunction with circling minima.

height above landing
The height above a designated helicopter landing area used for helicopter instrument approach procedures.

height above touchdown
The decision height (DH) or minimum decision altitude (MDA) above the highest elevation in the touchdown zone, for use in straight-in minima.

height ring
The ground return from a vertical sidelobe in a weather radar gives a bright height ring on the plan position indicator (PPI) centred on the origin.

helicopter
A rotorcraft that, for its horizontal motion, depends principally on its engine-driven rotors.

Helicopter Club of Great Britain
Objectives: An association for the sporting and social use of helicopters, and the FAI body in the United Kingdom for helicopter events.

heliport
An area of land, water or structure used or intended to be used for the landing and takeoff of helicopters.

hemispheric rule
An FAA requirement, that for all IFR flights, and for VFR flights from 3000 ft above the surface to FL 290, the cruising altitude is determined by the magnetic direction of flight.

henry
The inductance of a closed circuit in which an electromotive force of 1 volt is produced when the electric current in the circuit varies uniformly at a rate of 1 ampere per second.

hertz
Unit of frequency of 1 cycle per second; kilohertz (kHz) = 1000 cycles per second; megahertz (MHz) = 1 000 000 cycles per second.

heterodyne
The beat effect, used in radio receivers, made by combining two waves: a carrier wave and a wave of a different frequency.

high
An area of high barometric pressure, with its attendant system of winds; an anticyclone; a high pressure system.

high frequency
Radio frequencies from 3 MHz to 30 MHz (3000 kHz to 3 000 000 kHz).

high frequency single sideband
Voice modulation of aeronautical (HF) communications using high frequencies (3–30 MHz) with carrier suppressed to improve reception over a greater distance; reception is also improved by reducing side interference and signal fading that may be encountered using the standard AM signals.

high speed exit   See HIGH SPEED TAXIWAY.

high speed taxiway
A long radius taxiway designed and provided with lighting or marking to define the path of aircraft, travelling at high speed (up to 60 knots), from the runway centre to a point on the centre of the taxiway. The high speed taxiway is designed to expedite aircraft turning off the runway after landing, thus reducing runway occupancy time.

high speed turn-off   See HIGH SPEED TAXIWAY.

hinge moment
The magnitude of the aerodynamic twisting force on the hinge line of a control surface.

hoar frost   See FROST.

hold/holding procedure
A predetermined manoeuvre which keeps aircraft within a specified airspace while awaiting further clearance from air traffic control. Also used during ground operations to keep aircraft within a specified area or at a specified point while awaiting further clearance from air traffic control.

holding fix
A specified fix identifiable to a pilot by navaids or visual reference to the ground, used as a reference point in establishing and maintaining the position of an aircraft while holding. See also FIX and HOLD/HOLDING PROCEDURE.

holding point
(1) A specified location, identified by visual or other means, in the vicinity of which the position of an aircraft in flight is maintained in accordance with air traffic control clearances.
(2) Used during ground operations to keep aircraft within a specified area or at a specific point while awaiting further clearance from air traffic control.

homing
Flight towards or away from a radio station, using a radio direction finder as a primary means of maintaining directions.

hole
Electronics – a vacant electron energy state that is manifested as a charge defect in a crystalline solid, the defect behaving as a positive charge carrier with a charge magnitude equal to that of the electron.

homing guidance
The instrument system which enables a missile to find and destroy a target.

horizon, gyro   See GYRO HORIZON.

horizontal plane
The plane containing the longitudinal axis and perpendicular to the plane of symmetry of the aircraft.

horizontal situation indicator
An instrument displaying information from radio navigation aids, in the form of deviation signals which, in the case of VOR, relate to a course selected on the same instrument.

horn balance
That part of an aircraft control surface which is arranged to be ahead of the hinge line. Its purpose is to reduce the force required by the pilot to move the control surface.

horsepower
550 foot-pound per second = 745.7 watts.

hot mic
A microphone which is permanently live irrespective of crew-operated switch positions.

humidity
Absolute humidity is the actual percentage of water vapour in the air. Relative humidity is the saturation percentage at that temperature.

hurricane
A tropical cyclone in the western hemisphere with winds in excess of 65 knots or 120 km/h.

hydraulics
Devices which employ the incompressibility of fluids so that a force applied in one direction in one place can be transferred along tubes and applied in a different direction in a different place (e.g. car brakes); often with mechanical advantage, i.e. a small force over a long distance being changed into a large force over a short distance (e.g. car jacks).

hydrocarbons
Substances that contain hydrogen and carbon only, e.g. coal, oils, fuels, inflammable gases.

hydrogen (H)
Atomic number 1, normally the nucleus contains one proton and no neutron. However, two of its heavy isotopes (deuterium and tritium) are used in nuclear devices.

hydrometeor
A general term for particles of liquid water or ice such as rain, fog, frost, etc. formed by modification of water vapour in the atmosphere; water or ice particles lifted from the earth by the wind, such as sea spray or blowing fog.

hydrometer
An instrument which, when floated in liquid, will measure its relative density.

hydrophilic
A substance which has an affinity for water, e.g. which will absorb humidity from the air.

hydrosphere
The water of the oceans and seas: 85.8% oxygen, 10.7% hydrogen, 2.1% chlorine, 1.1% sodium, 0.14% magnesium. It contains many other elements, but all account for less than 0.05%. It covers more than two thirds of the earth's surface.

hygrograph
The record produced by a continuous-recording hygrometer.

hygrometer
An instrument for measuring the water vapour content of the air.

hygroscopic
Readily absorbing moisture, as from the atmosphere.

hyperbolic navigation
A means of navigation using a co-ordinate system of hyperbolic lines defined by ground-based radio transmitters.

hypergolic
Substances in rocket fuels which ignite when coming into contact with each other.

hypermetropia
Long sight; opposite of myopia.

hypersonic
Mach 5 and above.

hypotenuse
The longest side in a right angled triangle.

hypsometric tints
A succession of shades or colour gradations used to depict ranges of elevation.

hysteresis
When the results lag behind the cause, e.g. when strain lags behind stress in metals, it is less when stress is increasing and more when it is decreasing. This phenomenon also occurs with magnetism induced by an alternating current.

I

ice
Solid water at 0°C or less. The latent heat of fusion of ice is 80 calories per gramme and the specific heat is 0.55 calories per gramme.

ice crystals
A type of precipitation composed of unbranched crystals in the form of needles, columns or plates. Usually having a very slight downward motion, may fall from a cloudless sky.

ice fog
A type of fog composed of minute suspended particles of ice. Occurs at very low temperatures and may cause halo phenomena.

ice needles
Thin crystals or shafts of ice, so light that they seem to be suspended in the air.

ice pellets
Small, transparent or translucent, round or irregularly shaped pellets of ice. They may be either hard grains that rebound on striking a hard surface, or pellets of snow encased in ice.

icing
In general, any deposit of ice forming on an object. Airframe icing accumulation is reported in four categories:

(1) Trace. Ice becomes perceptible. Rate of accumulation slightly greater than the rate of sublimation. It is not hazardous even though de-icing/anti-icing equipment is not utilised, unless encountered for more than one hour.
(2) Light. The rate of accumulation might create a problem if flight in this environment exceeds one hour. Occasional use of de-icing/anti-icing equipment removes/prevents accumulation. It does not present a problem if de-icing/anti-icing equipment is used.
(3) Moderate. The rate of accumulation is such that even short encounters become potentially hazardous and use of de-icing/anti-icing equipment, or diversion, is necessary.
(4) Severe. The rate of accumulation is such that de-icing/anti-icing equipment fails to reduce or control the hazard. Immediate diversion is necessary.

See also CLEAR ICING, GLAZE and RIME.

ident
A request for a pilot to activate the aircraft transponder identification feature. This will help the controller to confirm an aircraft identity or to identify an aircraft.

ident feature
The special feature in the air traffic control radar beacon system (ATCRBS) equipment. It is used to distinguish immediately one displayed beacon target from other beacon targets. See also IDENT.

identification beacon
An aeronautical beacon emitting a coded signal by means of which a particular point of reference can be identified.

identification manoeuvre
A change of heading, requested by ATC, to establish positive identification of an aircraft by radar.

idle cut-off
A component of the manual mixture control of a piston engine. It is provided so that the engine can be stopped without leaving a combustible mixture in the induction passages, cylinders and exhaust system.

idle thrust
The jet thrust obtained with the engine power control lever set at the stop for the least thrust position at which it can be placed.

idling
In relation to an engine, operating speed with the throttle closed or nearly so.

if no transmission received by (time)
Used by ATC in radar approaches to prefix procedures that should be followed by the pilot in event of lost communications. See also LOST COMMUNICATIONS.

IFR aircraft
An aircraft conducting flight in accordance with instrument flight rules.

IFR conditions
Weather conditions below the minimum for flight under visual flight rules.

IFR flight   See IFR AIRCRAFT.

IFR landing minima   See LANDING MINIMA.

IFR military training routes [US]
Routes used by the Department of Defense and associated Reserve and Air Guard units for the purpose of conducting low-altitude navigation and tactical training in both IFR and VFR weather conditions below 10 000 ft msl at airspeeds in excess of 250 knots IAS.

igniter plug
An electric device used for the starting of gas turbine engines.

ignition harness
HT cables which are screened in a complete assembly for the purpose of serving electrical power to the spark plugs of a piston engine, without causing interference to the aircraft radio system.

ignition point
The minimum temperature at which a fuel/air mixture can be ignited by a spark or other means.

image
A real image formed by a concave mirror or a convex lens that can be projected onto a screen.

image converter
Transforms images made by non-visible radiation into visible images.

immediately
Used by ATC when compliance with such action is required to avoid an imminent situation.

impedance
The extent to which the flow of alternating current at a given frequency is restricted, and represented by the ratio of root mean square (rms) values of voltage and current. Combines resistance, capacity and inductive reactance.

imperial units
The old British system of measurements using inches, feet, yards, ounces, pounds, pints, quarts, gallons, etc.

implosion
The rapid action of forces towards the centre; a structure collapsing inwards.

impulse
A force of short duration.

impulse magneto
A magneto with a drive mechanism which incorporates stops and a spring loaded coupling to give a sudden rotation and strong spark during starting.

incandescence
Light produced by high temperatures.

incidence
The angle of incidence of a wing is the angle (of attack) with which the root joins the fuselage; the angle of incidence of, say, a ray of light is the angle at which it meets the surface measured from the normal (vertical).

inclinometer
An instrument for measuring the angle of inclination of an aircraft to the horizontal; an instrument for measuring the dip in the earth's magnetic field.

incremental sensitivity
The increment of receiver indicator current per unit change of receiver antenna displacement from the nominal course line or nominal ILS glide path.

indefinite ceiling
A ceiling classification denoting vertical visibility into a surface-based obscuration.

indicated airspeed
The speed of an aircraft as shown on its pitot static airspeed indicator calibrated to reflect standard atmosphere adiabatic compressible flow at sea level, uncorrected for airspeed system errors.

indicated altitude
The altitude above mean sea level indicated on a pressure altimeter set at current local altimeter setting.

indicated ILS glide path
The locus of points in the vertical plane containing the runway centreline at which the receiver indicator deflection is zero.

indicated ILS glide path angle
The angle above the horizontal plane of the indicated ILS glide path.

indicated ILS glide path sector
The sector containing the indicated ILS glide path in which the receiver indicator deflection remains within full-scale values.

indicated slant course line
The line formed at the intersection of the indicated course surface and the plane of the nominal ILS glide path.

induced drag
The drag created by the production of lift from an aerofoil.

induced velocity
Air circulation around an aerofoil and downwash usually considered as proportional to lift. In helicopters, the downward air velocity generated in the process of developing upward rotor thrust.

inductance
The property of an element or circuit which, when carrying a current, is characterised by the formation of a magnetic field and the storage of magnetic energy.

induction coil
A larger current with a lower voltage in a primary winding induces a smaller current with a higher voltage in a secondary winding.

inert gases
Helium, neon, argon, krypton, xenon and radon. All chemically inactive even at very high temperatures. Also called the noble gases.

inertia
A property of mass which resists changing its uniform motion in a straight line, or its state of rest if it is motionless.

inertial navigation system
A non-radio navigation aid which computes the aircraft's position by dead reckoning using the measured acceleration of an airborne gyro stabilised platform.

inflight refuelling   See AERIAL REFUELLING.

inflow
The component of perpendicular air velocity through the rotor disc of a helicopter.

information request [US]
A request originated by an FSS for information concerning an overdue VFR aircraft.

infra-red radiation
Invisible heat radiation (radiant heat) which can pass through clouds and haze. Used for special high altitude photography.

infrasonic
With a frequency of less than 20 Hz, i.e. below the range of audible sounds.

initial approach
That part of an instrument approach procedure consisting of the approach to the first navigational facility associated with the procedure, or to a predetermined fix.

initial approach area
An area of defined width lying between the last preceding navigational fix or dead reckoning position and either the facility to be used for making an instrument approach or a point associated with such a facility that is used for demarcating the termination of initial approach.

initial approach fix
The fix(es) depicted on instrument approach procedure charts that identify the beginning of the initial approach segment(s). See also FIX, SEGMENTS OF AN INSTRUMENT APPROACH PROCEDURE.

initial approach (IA) mode
The condition of DME/P operation which supports those flight operations outside the final approach region.

initial approach segment
That segment of an instrument approach procedure between the initial approach fix and the intermediate approach fix or, where applicable, the final approach fix or point.

inlet duct
The passage through which air flows to the carburettor.

inner marker
A marker beacon used with an ILS (CAT II) precision approach located between the middle marker and the end of the ILS runway, transmitting a radiation pattern keyed at six dots per second and indicating to the pilot, both aurally and visually, that he is at the designated decision height (DH), normally 100 ft above the touchdown zone elevation, on the ILS CAT II approach. It also marks progress during a CAT III approach. See also INSTRUMENT LANDING SYSTEMS.

inner marker beacon   See INNER MARKER.

insolation
Exposure to sunlight.

insoluble
Cannot be dissolved.

instability
A state in which the vertical distribution of temperature is such that an air particle, if given either an upward or a downward impulse, will tend to move away with increasing speed from its original level.

Institution of Analysts and Programmers
Objectives: A non-profit making organisation for the advancement and greater recognition of the profession of systems analysis and computer programming and for the encouragement of the wider use of analysis and programming in technical undertakings and their associated activities.

Institution of Mechanical Engineers
Objectives: To promote standards of excellence in British mechanical engineering and a high level of professional development, competence and conduct among aspiring and practising mechanical engineers. The Aerospace Industries Division is specifically responsible for the furtherance of these objectives amongst engineers in the various aerospace related industries.

instrument approach procedure
A series of predetermined manoeuvres by reference to flight instruments with specified protection from obstacles from the initial approach fix or, where applicable, from the beginning of a defined arrival route, to a point from which a landing can be completed, and thereafter, if a landing is not completed, to a position at which holding or en-route obstacle clearance criteria apply.

instrument approach procedure [ICAO]
A series of predetermined manoeuvres for the orderly transfer of an aircraft under instrument flight conditions from the beginning of the initial approach to a landing or to a point from which a landing may be made visually.

instrument approach runway
A runway intended for the operation of aircraft using non-visual aids providing at least directional guidance in azimuth adequate for a straight-in approach.

instrument approach strip
An area of specified dimensions which encloses an instrument runway.

instrument flight rules
Rules governing the procedures for conducting instrument flight; a term used by pilots and controllers to indicate type of flight plan. See also INSTRUMENT METEOROLOGICAL CONDITIONS, VISUAL FLIGHT RULES and VISUAL METEOROLOGICAL CONDITIONS.

instrument flight rules [ICAO]
A set of rules governing the conduct of flight under instrument meteorological conditions.

instrument flight time
Time during which a pilot is piloting an aircraft solely by reference to instruments and without external reference points.

instrument ground time
Time during which a pilot is practising, on the ground, simulated instrument flight on a synthetic flight trainer approved by the Licensing Authority.

instrument landing systems [ICAO]
Aids for an instrument approach to an airport, consisting of a VHF localiser beam for lateral guidance; a UHF glide slope, for vertical guidance; a VHF outer marker and middle marker, and sometimes an inner marker, for determining position along the approach path. A compass locator is positioned at the outer marker and, in special cases, at the middle marker. DME is often available also. ILS approach categories are as follows (operational performance):

(1) ILS category I. An ILS approach procedure that provides for approach to a height above touchdown of not less than 200 ft and with runway visual range of not less than 800 m.
(2) ILS category II. An ILS approach procedure that provides for approach to a height above touchdown of not less than 100 ft and with runway visual range of not less than 400 m.
(3) ILS category III:

   IIIA. An ILS approach procedure that provides for approach without a decision height minimum and with runway visual range of not less than 200 m.
   IIIB. An ILS approach procedure that provides for approach without a decision height minimum and with runway visual range of not less than 50 m.
   IIIC. An ILS approach procedure that provides for approach without a decision height minimum and without any runway visual range minimum.

(4) ILS continuity of service. That quality which relates to the rarity of radiated signal interruptions during any approach. The level of continuity of service of the localiser or the glide path is expressed in terms of the probability of not losing the radiated guidance signals.
(5) ILS glide path. The locus of points in the vertical plane containing the runway centreline at which the receiver indicator deflection is zero.
(6) ILS glide path angle. The angle above the horizontal plane of the indicated ILS glide path.
(7) ILS glide path bend. An ILS glide path bend is an aberration of the ILS glide path with respect to its nominal position.
(8) ILS glide path sector. The sector containing the indicated ILS glide path in which the receiver indicator deflection remains within full-scale values.
   Note: The ILS glide path sector is located in the vertical plane containing the runway centreline and is divided by the radiated glide path in two parts called upper sector and lower sector, referring respectively to the sectors above and below the glide path.
(9) ILS integrity. That quality which relates to the trust which can be placed in the correctness of the information supplied by the facility. The level of integrity of the localiser or the glide path is expressed in terms of the probability of not radiating false guidance signals.
(10) ILS point A. A point on the ILS glide path measured along the extended

runway centreline in the approach direction a distance of 7.5 km (4 NM) from the threshold.

(11) ILS point B. A point on the ILS glide path measured along the extended runway centreline in the approach direction a distance of 1050 m from the threshold.

(12) ILS point C. A point through which the downward extended straight portion of the nominal ILS glide path passes at the height of 30 m above the horizontal plane containing the threshold.

(13) ILS point D. A point 4 m above the runway centreline and 900 m from the threshold in the direction of the localiser.

(14) ILS point E. A point 4 m above the runway centreline and 600 m from the stop end of the runway in the direction of the threshold.

(15) ILS reference datum (point T). A point at a specified height located vertically above the intersection of the runway centreline and the threshold and through which the downward extended straight portion of the ILS glide path passes.

instrument meteorological conditions

Meteorological conditions expressed in terms of visibility and distance from cloud and ceiling which is less than the minima specified for visual meteorological conditions. See also INSTRUMENT FLIGHT RULES, VISUAL FLIGHT RULES and VISUAL METEOROLOGICAL CONDITIONS.

instrument rating

A qualification enabling a pilot to operate in instrument weather conditions and in certain types of airspace.

instrument runway

A runway equipped with electronic visual navigation aids for which a precision or non-precision approach procedure having straight-in landing minimums has been approved.

instrument runway [ICAO]

A runway intended for the operation of aircraft using non-visual aids and comprising:

(1) Instrument approach runway. An instrument runway served by a non-visual aid providing at least directional guidance adequate for a straight-in approach.

(2) Precision approach runway, category I. An instrument runway served by ILS or GCA approach aids and visual aids intended for operations down to 60 m (200 ft) decision height and down to an RVR of the order of 800 m (2600 ft).

(3) Precision approach runway, category II. An instrument runway served by ILS and visual aids intended for operations down to 30 m (100 ft) decision height and down to an RVR of the order of 400 m (1200 ft).

(4) Precision approach runway, category III. An instrument runway served by ILS (no decision height being applicable) and:

    (a) By visual aids intended for operations down to an RVR of the order of 200 m (700 ft).

    (b) By visual aids intended for operations down to an RVR of the order of 50 m (150 ft).

    (c) Intended for operations without reliance on visual reference.

insulator
A substance that will not conduct either heat or electricity or both.

integrated circuit
A silicon chip of semiconductors incorporating a whole system. Used in computers.

integrated European region automated AIS system [ICAO]
An aeronautical information service (AIS) developed to provide an interface with the individual automated AIS systems of European countries.

integrated flight system
An integrated flight system consists of electronic components which compute and indicate the aircraft attitude required to attain and maintain a preselected flight condition. 'Command' indicators tell the pilot in which direction and how much to change aircraft attitude to achieve the desired result. The computed command indications relieve the pilot of many of the mental calculations.

integrated flight training
A system of flight instruction under which, from the beginning, attitude control is taught both by visual references outside the aircraft, and by reference to the primary aircraft instruments. Instruction in the use of both systems of reference proceeds concurrently.

intensity modulation
Variation of the velocity of the electron beam in a cathode ray tube so as to cause a corresponding variation in intensity of the brightness of the display.

interception
The problem of determining the direction of flight, the direction and speed of relative movement, and the time required to intercept another craft in motion.

interface
The point at which two parts of a system or two systems meet.

interference
The destruction or distortion of one wave by another, or one broadcast by another.

interferometer
An antenna array, together with phase discriminators, capable of measuring the direction of arrival of an electromagnetic wave.

interlace
Time multiplexing of modes of interrogation in a secondary radar system; in particular, modes A and C may be interlaced in SSR.

intermediate approach
That part of an instrument approach procedure from the first arrival at the first navigational facility of pre-determined fix, to the beginning of final approach.

intermediate approach segment   See SEGMENTS OF AN INSTRUMENT APPROACH PROCEDURE.

intermediate fix
The fix that identifies the beginning of the intermediate approach segment of an instrument approach procedure. The fix is not normally identified on the instrument approach chart as an intermediate approach fix (IF). See also SEGMENTS OF AN INSTRUMENT APPROACH PROCEDURE.

internal combustion engine
An engine in which the fuel is burned inside the working parts (piston engines, turboprops, jets).

internal stress
Stress inside a solid caused by temperature change, heat treatment or deformation.

International Airline Passengers Association
Objectives: To promote safety in airline travel, improve passenger handling, comfort and facilities, and work with the airlines to achieve these aims.

international airport [ICAO]
Any airport designated by the contracting State in whose territory it is situated as an airport of entry and departure for international air traffic, where the formalities incident to customs, immigration, public health, animal and plant quarantine and similar procedures are carried out.

international airport [US]
Relating to international flights:

(1) An airport of entry that has been designated by the Secretary of Treasury or Commissioner of Customs as an international airport for customs service.
(2) A landing rights airport at which specific permission to land must be obtained from customs authorities in advance of contemplated use.
(3) Airports designated under the Convention on International Civil Aviation as an airport for use by international commercial air transport and/or international general aviation. (Refer to Airport/Facility Directory.)

international air service
An air service which passes through the airspace over the territory of more than one State.

International Air Transport Association
Objectives: To promote safe, regular and economical air transport; to provide means for collaboration among international air transport companies; to co-operate with ICAO and other international organisations.

International Business Aviation Council
Objectives: Involved in activities which ensure that the needs and interests of business aviation on an international scale are clearly presented to, and understood by, those national and international authorities and organisations which influence the safety, efficiency or economic use of business aircraft operating internationally. This is to ensure ever widening recognition of the fact that international operations conducted by business aircraft are of primary importance to the economy and well-being of all nations.

International Civil Aviation Organisation

A specialised agency of the United Nations whose objective is to develop the principles and techniques of international air navigation and to foster planning and development of international civil air transport. ICAO has a sovereign body, the Assembly, and a governing body, the Council. The Assembly meets at least once in three years and is convened by the Council. Each Contracting State is entitled to one vote and decisions of the Assembly are taken by a majority of the votes cast except when otherwise provided in the Convention. At this session the complete work of the Organisation in the technical, economic, legal and technical assistance fields is reviewed in detail and guidance given to the other bodies of ICAO for their future work. The Council is a permanent body responsible to the Assembly and is composed of 33 Contracting States elected by the Assembly for a three-year term. In the election, adequate representation is given to States of chief importance in air transport, States not otherwise included which make the largest contribution to the provision of facilities for civil air navigation and States not otherwise included whose designation will ensure that all the major geographic areas of the world are represented on the Council. The Council, the Air Navigation Commission, the Air Transport Committee, the Committee on Joint Support of Air Navigation Services and the Finance Committee provide the continuing direction of the work of the Organisation. One of the major duties of the Council is to adopt international standards and recommended practices and to incorporate these as Annexes to the Convention on International Civil Aviation. The Council may act as an arbiter between Member States on matters concerning aviation and implementation of the Convention; it may investigate any situation which presents avoidable obstacles to the development of international air navigation and, in general, it may take whatever steps are necessary to maintain the safety and regularity of operation of international air transport.

International Council of Aircraft Owner and Pilot Associations

A non-profit-making federation of (in 1989) 35 autonomous, non-governmental, national general aviation organisations. The IAOPA has represented international general aviation since 1962. The combined total of individuals represented by these constituent member groups of IAOPA is over 400 000 pilots, who fly aircraft for aerial work, business and personal transportation. Objectives: To facilitate the movement of general aviation activities internationally, for peaceful purposes in order to develop friendship and understanding among the peoples of the world and to increase the utility of the general aviation sector as a means of personal and business transportation, and aerial work, bringing benefits to the world's communities.

international date line

An imaginary line, sited approximately along the 180° meridian.

International Federation of Air Line Pilots' Associations

Objectives: The Federation's basic aim is the development of a safe and orderly system of air transportation and the protection of the interests of airline pilots.

Consists of many national pilot associations totalling approximately 60 000 pilots around the world.

International Federation of Airworthiness
Objectives: To provide a forum for the exchange of international experience and ideas on all matters concerned with airworthiness, be it maintenance, design, or operations. Membership consists of a large number of organisations, including manufacturers, air worthiness authorities, airlines, fixed base operators, technical training schools and professional societies.

International Flight Information Manual [US]
A publication designed primarily as a pilot's preflight planning guide for flights into foreign airspace.

International Maritime Satellite Organisation
Objectives: To provide satellite communications facilities for maritime, aeronautical and other mobile applications; and satellite services for aircraft, including passenger telephone calls, which began on a trial and demonstration basis in 1987.

International NOTAM Office
An office designated by a State for the exchange of NOTAM internationally.

International Standards and Recommended Practices
These are adopted by the council of ICAO in accordance with Articles 54, 37 and 90 of the Convention on International Civil Aviation and are designated, for convenience, as Annexes to the Convention. The uniform application by Contracting States of the specifications contained in the International Standards is recognised as necessary for the safety or regularity of international air navigation, while the uniform application of the specifications in the Recommended Practices is regarded as desirable in the interests of safety, regularity or efficiency of international air navigation. Knowledge of any differences between the national regulations or practices of a State and those established by an International Standard is essential to the safety or regularity of international air navigation. In the event of non-compliance with an International Standard, a State has, in fact, an obligation under Article 38 of the Convention to notify the Council of any differences. Knowledge of differences from Recommended Practices may also be important for the safety of air navigation and, although the Convention does not impose any obligation with regard thereto, the Council has invited Contracting States to notify such differences in addition to those relating to International Standards.

International telecommunications service
A telecommunications service between offices or stations of different States, or between mobile stations which are not in the same State, or are subject to different States.

International Union of Aviation Insurers
An official body which is able to speak on behalf of international aviation insurance interests, to provide a central office for the circulation of information between members, and to assist in, and provide, for the better understanding and conduct of international aviation insurance.

interpilot air-to-air communication
Two-way communication on a designated air-to-air channel to enable aircraft engaged in flights over remote and oceanic areas out of range of VHF ground stations to exchange necessary operational information and to facilitate the resolution of operational problems.

interplanetary space
The space between the planets inside our solar system.

Inter Range Instrumentation Group
Affiliated to the United States Defense Department, in 1948 this group set standards for instrumentation recording, which have been revised periodically and which are still recognised.

interrogator
The ground-based surveillance radar beacon transmitter–receiver that normally scans in synchronism with a primary radar, transmitting discrete radio signals that repetitiously requests all transponders, on the mode being used, to reply. The replies received are mixed with the primary radar returns and displayed on the same plan position indicator (radar scope). Also applied to the airborne element of the TACAN/DME system. See also TRANSPONDER.

interscan
(1) The period between the end of one scan of a timebase on a radar display and the commencement of the next.
(2) The name of the Australian contender for the microwave landing system international standard. This operates on time reference scanning beam principles and the proposal was eventually combined with that from the USA to form the preferred standard.

intersecting runways
Two or more runways that cross or meet within their lengths. See also INTERSECTION.

intersection
(1) A point defined by any combination of courses, radials or bearings of two or more navigational aids.
(2) Used to describe the point where two runways cross, a taxiway and a runway cross, or two taxiways cross.

intersection departure
A takeoff or proposed takeoff on a runway from an intersection. See also INTERSECTION.

intersection takeoff See INTERSECTION DEPARTURE.

interstate air transportation [US]
The carriage by aircraft of persons or property as a common carrier for compensation or hire, or the carriage of mail by aircraft, in commerce:

(1) Between a place in a State or the District of Columbia and another place in another State or the District of Columbia;

(2) Between places in the same State through the airspace of any place outside that State;

(3) Between places in the same possession of the United States; whether that commerce moves wholly by aircraft or partly by aircraft and partly by other forms of transportation.

interstellar space
The space between the stars in our galaxy (the Milky Way).

intertropical convergence zone
The boundary zone between the trade wind system of the northern and southern hemispheres; it is characterised in maritime climates by showery precipitation with cumulonimbus clouds sometimes extending to great heights.

intertropical front
The boundary between the trade wind systems of the Northern and Southern Hemispheres. It manifests itself as a fairly broad zone of transition commonly known as the doldrums.

intrastate air transportation [US]
The carriage of persons or property as a common carrier for compensation or hire, by turbojet-powered aircraft capable of carrying 30 or more persons, wholly within the same state of the United States.

invar
A nickel/iron alloy with a coefficient of linear expansion which is only about 8% that of ordinary steel. It is used in instruments to prevent inaccuracies due to temperature change.

inverse square law
The effect of one body on another decreased as the square of the distance between them. This law applies to gravitation, magnetism and illumination.

inversion
A layer in which the temperature increases with altitude.

ion engine
For rocket propulsion in space. The thrust derives from a jet of ions which repel and are repelled by the body of the rocket.

Ionosphere
That part of the earth's atmosphere which is above 50 km from the surface; so called because of the ionisation caused by ultraviolet and X-rays from the sun. It is used for reflecting sky waves (radio frequencies) from transmitting stations, but it hinders radio astronomy by reflecting most of the radiation from space. It is divided into three regions: the D-region from 50 km to 90 km; the E-region or Heaviside-Kennelly layer from 90 km to 150 km; and the Appleton layer, above 150 km.

Ionospheric refraction
The change in the propagation speed of a signal as it passes through the Ionosphere.

ions
Atoms which have either gained or lost electrons. Those which have gained electrons have become negative but are called 'anions' because they are attracted to the anode. Those which have lost electrons have become positive but are called 'cations' because they are attracted to the cathode.

iron (Fe)
Atomic number 26, relative density 7.86, melting point 1535°C; magnetic.

isallobar
On a weather chart, lines joining places of equal barometric pressure change.

iso
Equal, identical or similar.

isobars
Lines on maps which join points of equal barometric pressure.

isoclines
Lines on maps which join points of equal magnetic dip.

isodynamic lines
On maps of the earth's magnetic field these lines join points of equal strength.

iso echo
In radar circuitry, a circuit that reserves signal strength above a specified intensity level, thus causing a void on the scope in the most intense portion of an echo when maximum intensity is greater than the specified level.

isogonal lines
Lines joining points of equal magnetic variation.

isogriv
A line on a map or chart which joins points of equal angular difference between the north of navigation grid and magnetic north.

isoheight
On a weather chart, a line of equal height; same as a contour.

isoline
A line of equal value of a variable quantity, i.e. an isoline of temperature is an isotherm, etc. See also ISOBAR, ISOTACH, etc.

iso-octane
Hydrocarbon of paraffin series which is used as a reference measure for determing the anti-knock rating of fuels.

isosceles triangle
A triangle with two sides of equal length and two angles equal.

isoshear
A line of equal wind shear.

isotach
A line of equal or constant wind speed.

isothermal lines
Lines passing through points with the same temperature.

isotropic
With the same uniform properties in all directions.

J

jamming
Electronic or mechanical interference that may disrupt the display of aircraft on radar or the transmission/reception of radio communications/navigation.

jet blast
Jet engine exhaust (thrust stream turbulence). See also WAKE TURBULENCE.

jet engine
A gas turbine producing a foward thrust by accelerating exhaust gases rearward: reaction propulsion. About 90% efficient at 1100 knots but a 'gas-guzzler' at slow speeds or low altitude.

jet route [US]
A route designed to serve aircraft operations from 18 000 ft amsl up to and including flight level 450. The routes are referred to as 'J' routes with numbering to identify the designated route, e.g. J 105. See also ROUTE.

jet stream
A narrow meandering stream of winds with speeds of 150 knots and greater, embedded in the normal wind flow in the high troposphere.

jettisoning of external stores
Airborne release of external stores, e.g. tiptanks, ordnances. See also FUEL DUMPING.

joining point
The point at which an aircraft enters or is expected to enter a control area from uncontrolled airspace.

Joint Airmiss Working Group [UK]
A group consisting of representatives of civil and military aviation, which reviews all reported airmisses which occur in the UK airspace and the airspace surrounding the British Isles. In their review of airmiss reports the members of the JAWG assess the degree of risk inherent in each occurence, determine the cause, take note of remedial action already taken and, when appropriate, record their comments and recommendations. The degree of risk is categorised as follows:

(1) Category A. Actual risk of collision.
(2) Category B. Possible risk of collision.
(3) Category C. Other reports with no assessed risk of collision.

Joint Airport Committee of Local Authorities [UK]
Objectives: The successor organisation to the Municipal Airports Committee originally set up in 1935. JACOLA is an elected member organisation representing

the interests of all local authority owned airports in England, Scotland and Wales. Maintains links with Central Government, the EEC, similar organisations in other countries and all organisations concerned with the civil aviation industry.

Joint Aviation Authorities

Formerly the Joint Airworthiness Requirements or JAR systems Group. The JAR process was developed to standardise aircraft and equipment certification between European countries and to harmonise where possible the European and FAA requirements. These objectives have subsequently been expanded to include maintenance requirements. A number of the participating European countries have also signed a Memorandum of Understanding which commits each authority to complete all the airworthiness codes, enables the authorities to adopt them as their only standards for new products and extends the work to maintenance and operational matters. The JAA is removing the burden of multiple certifications and different standards, is creating common or mutually acceptable maintenance systems and relies on the principle of worksharing between countries. Small international teams are used in the joint certification procedure. For a European product a team is selected and approved by all countries with two specialists in each discipline relevant to the project. One member of each discipline is from the domestic authority. It is an accepted principle that the team is selected on the basis of knowledge and experience, not to achieve national 'balance'. For the product of an outside country there is no domestic authority and both specialists are selected from JAA personnel.

joule

The SI unit of energy = 1 watt second = 1 newton metre = 10^7 ergs = 0.24 calories.

Joule's laws

(1) The internal energy of a gas at constant temperature is independent of its volume.

(2) The heat produced by an electric current passing through a resistance = I^2Rt or amps^2 × ohms × seconds = joules.

K

katabatic wind
Any wind blowing downslope. See also FALL WIND and FOEHN.

Kelvin temperature scale
A temperature scale with 0° equal to the temperature at which all molecular motion ceases, i.e. absolute zero (0°K = −273°C). The kelvin degree is identical to the Celsius degree. Hence at standard sea level pressure, the melting point of water is 273°K and the boiling point 373°K.

Kepler's laws
(1) The paths of the planets around the sun are ellipses with the sun at one focus.
(2) The radius vector joining each planet to the sun covers equal areas during equal periods of time.
(3) The square of the planet's year divided by the cube of its average distance from the sun is the same for all planets in our solar system.

kerosine
Paraffin oil. It is a variable mixture of hydrocarbons with boiling points from 150°C–300°C. Also spelt kerosene.

key down time
The time during which a dot or a dash of a Morse character is being transmitted.

kilocycle
A frequency of 1000 cycles per second, or one kilohertz.

kilogram (Kg)
The SI unit of mass = 2.20462 lb. Also spelt kilogramme.

kilohertz (kHz)
1000 hertz.

kilometre (km)
1000 metres = 1094 yards = 0.6214 mile.

kilowatt (kW)
1000 watts of power.

kilowatt-hour (kWh)
3 600 000 watt seconds (joules) = 1 commercial unit of electricity consumption.

kinematics
The measurement and description of motion.

kinetic energy
The kinetic energy of a mass (m) moving at a velocity (v)

$v = \frac{1}{2}mv^2$

or $\frac{1}{2}kg\left(\frac{m}{s}\right)^2$ = joules where (m) = metres and (s) = seconds.

The kinetic energy of rotation of a body with a moment of inertia I and an angular velocity omega (ω), is $\frac{1}{2}I\omega^2$

knocking
The sound of loud explosions in the cylinders of internal combustion engines due to over compression.

knot
A unit of speed equal to one nautical mile per hour.

known traffic
With respect to ATC clearances, aircraft whose altitude, position and intentions are known to ATC.

kovar
An alloy used for glass-to-metal seals. Its coefficient of expansion is almost the same as glass and it is much cheaper than platinum.

L

lading
The placing of cargo, mail, baggage or stores on board an aircraft to be carried on a flight, except such cargo, mail, baggage or stores as have been laden on a previous stage of the same through-flight.

lambert
The name of the chart projection used as a base for many aeronautical charts. Straight lines on this projection closely approximate great circles, and distances may be measured with a high degree of accuracy.

laminar flow
Streamlined (non-turbulent) flow which follows the surface without separating.

land breeze
A coastal breeze blowing from land to sea, caused by temperature difference when the sea surface is warmer than the adjacent land. Therefore it usually blows at night and alternates with a sea breeze, which blows in the opposite direction by day.

landing area
Any locality either of land or water, including airports and intermediate landing fields, that is used, or intended to be used, for the landing and takeoff of aircraft, whether or not facilities are provided for the shelter, servicing or for receiving or discharging passengers or cargo.

landing area [ICAO]
That part of the movement area intended for the landing and takeoff of aircraft.

landing direction indicator
A device that visually indicates the direction in which landings and takeoffs should be made. See also TETRAHEDRON.

landing gear extended speed
The maximum speed at which an aircraft can be safely flown with the landing gear extended.

landing gear operating speed
The maximum speed at which the landing gear can be safely extended or retracted.

landing minima
The minimum visibility prescribed for landing a civil aircraft while using an instrument approach procedure. The minimum applies with other limitations set forth by the particular aviation administration and/or aircraft operator with respect to the minimum descent altitude (MDA) or decision height (DH) prescribed in instrument approach procedures as follows:

(1) Straight-in landing minima. A statement of MDA and visibility or DH and visibility, required for straight-in landing on a specified runway.
(2) Circling minima. A statement of MDA and visibility required for the circle-to-land manoeuvre.

Descent below the established MDA or DH is not authorised during an approach unless the aircraft is in a position from which a normal approach to the runway of intended landing can be made, and adequate visual reference to required visual cues is maintained. See also CIRCLE-TO-LAND, DECISION HEIGHT, INSTRUMENT APPROACH PROCEDURE, MINIMUM DESCENT ALTITUDE, STRAIGHT-IN-LANDING and VISIBILITY.

landing roll
The distance from the point of touchdown to the point where the aircraft can be brought to a stop or exit the runway.

landing sequence
The order in which aircraft are positioned for landing.

landing surface
That part of the surface of an aerodrome which the aerodrome authority has declared available for the normal ground or water run of aircraft landing in a particular direction.

lapse rate
The rate of decrease of an atmospheric variable with height. Commonly refers to decrease of temperature with height.

large aircraft [US]
Aircraft of more than 12 500 lb, maximum certificated takeoff weight.

laser
Light amplification by simulated emission of radiation. A laser beam of light is coherent and very directional. It is produced in a transparent cylinder with a reflector at one end and a partial reflector at the other, or by the excitation of a cylindrical crystal. Used for cutting and welding metals, surgery and holography.

last assigned altitude
The last altitude/flight level assigned by ATC and acknowledged by the pilot. See also MAINTAIN.

latent heat
The heat lost or gained when a substance changes state without changing temperature, e.g. ice at 0°C to water at 0°C, or water at 100°C to steam at 100°C; the amount of heat absorbed (converted to kinetic energy) during the processes of change of liquid water to water vapour, or ice to liquid water, or the amount released during the reverse processes. Four basic classifications are:

(1) Latent heat of condensation. Heat released during change of water vapour to water.
(2) Latent heat of fusion. Heat released during change of water to ice or the amount absorbed in change of ice to water.

(3) Latent heat of sublimation. Heat released during change of water vapour to ice or the amount absorbed in the change of ice to water vapour.

(4) Latent heat of vaporisation. Heat absorbed in the change of water to water vapour; the negative of latent heat of condensation.

lateral datum line
An imaginary line through the aircraft's centre of gravity usually considered to be parallel to a line joining the wing tips:

lateral inversion
The left/right inversion of an image in a mirror.

lateral separation
The lateral spacing of aircraft at the same altitude by requiring operation on different routes or in different geographical locations. See also SEPARATION.

latitude
Measured in degrees north and south of the equator: the angular distance north and south. Parallels of latitude are the paths traced on the surface by radii.

layer
In reference to sky cover, clouds or other obscuring phenomena whose bases are approximately at the same level. The layer may be continuous or composed of detached elements. The term 'layer' does not imply that a clear space exists between the layers or that the clouds or obscuring phenomena composing them are of the same type.

L-band
The group of radio frequencies extending from 390 MHz to 1550 MHz. The global positioning system (satellite) carrier frequencies 1227.6 MHz and 1575.42 MHz are in the L-band.

leading edge
The forward edge of a streamlined body or aerofoil

leaving point
The point at which an aircraft leaves or is expected to leave a control area for uncontrolled airspace.

lee wave
Any stationary wave disturbance caused by a barrier in a fluid flow. In the atmosphere, when sufficient moisture is present, this wave will be evidenced by lenticular clouds to the lee of mountain barriers. Also called mountain wave or standing wave.

lenticular cloud
A type of cloud whose elements have the form of more or less isolated, generally smooth, lenses or almonds. These clouds appear most often in formations of orographic origin, the result of lee waves, in which case they remain nearly stationary with respect to the terrain (standing cloud), but they also occur in regions without marked orography.

lenticularis   See LENTICULAR CLOUD.

Lenz's law
When a conductor is moved across the lines of force in a magnetic field, the current induced in the conductor produces another magnetic field which opposes the motion.

level
A generic term relating to the vertical position of an aircraft in flight and meaning, variously, height, altitude or flight level.

level of free convection
The level at which a parcel of air lifted dry-adiabatically until saturated and moist-adiabatically thereafter would become warmer than its surroundings in a conditionally unstable atmosphere. See also ADIABATIC and CONDITIONAL STABILITY.

licence
A document issued by the Licensing Authority of a State permitting the holder to exercise specific privileges in relation to aircraft operations. In the United States this document is known as a Certificate. With respect to personnel licensing, the Transport Commission of the EC has defined the following:

(1) Licence. Any valid document, issued by a Member State, authorising personnel to exercise functions in civil aviation. This definition also includes ratings forming part of the licence.
(2) Rating. An authorisation entered on or associated with a licence, stating special conditions pertaining to such licence.
(3) Acceptance of licences. Any act of recognition or validation by a Member State of a licence or aspect of a licence issued by another Member State. The acceptance may be effected through the issue of a licence of its own.
(4) Recognition. The permission to use in one Member State a licence issued in another Member State for the purpose specified on the licence.
(5) Validation. The express indication by a Member State in a licence issued by another Member State that this licence can be used as one of its own for the purpose specified on the licence.

Licensing Authority [ICAO]
The Authority designated by a Contracting State as responsible for the licensing of aviation personnel. In the provisions of ICAO Annex 1, the Licensing Authority is deemed to have been given the following responsibilities by the Contracting State:

(1) assessment of an applicant's qualifications to hold a licence or rating;
(2) issue and endorsement of licences and ratings;
(3) designation and authorisation of approved persons;
(4) approval of training courses;
(5) approval of the use of synthetic flight trainers for their use in gaining experience or in demonstrating the skill required for the issue of a licence or rating;
(6) validation of licences issued by other ICAO Contracting States

lift
The sum of all the aerodynamic forces acting on an aircraft at right angles to the flight path. When the aircraft is in steady level flight the lift is equal and opposite to the weight of the aircraft.

lifting condensation level
The level at which a parcel of unsaturated air lifted dry-adiabatically would become saturated. See also CONVECTIVE CONDENSATION LEVEL and LEVEL OF FREE CONVECTION.

light
Electromagnetic radiations with wavelengths between 10^{-6} and 10^{-7} m. (Ultraviolet is shorter and infra-red is longer.) Velocity $= 2.997925 \times 10^8$ m/s.

light aircraft [ICAO]
For the purpose of weight turbulence categorisation, aircraft of 7000 kg or less are categorised as 'small'.

lighted airport
An airport where runway and obstruction lighting is available. See also AIRPORT LIGHTING.

light emitting diode
A semiconductor diode from which light is emitted, used to make digital displays on instruments.

light failure
A light shall be considered to have failed when for any reason the average intensity determined using the specified angles of beam elevation, toe-in and spread falls below 50% of the specified average intensity of a new light.

lighter-than-air aircraft
Aircraft than can rise and remain suspended by using contained gas weighing less than the air that is displaced by the gas.

light gun
A hand-held light signalling device that emits a brilliant narrow beam of white, green or red light as selected by the tower controller. The colour and type of light transmitted can be used to approve or disapprove anticipated pilot actions where radio communication is not available. The light gun (aldis) is used for controlling traffic operating in the vicinity of the airport and on the airport movement area.

lighting system reliability
The probability that the complete installation operates within the specified tolerances and that the system is operationally usable.

lightning
Generally, any and all forms of visible electrical discharge produced by a thunderstorm.

limited remote communications outlet
An unmanned satellite air/ground communications facility that may be associated with a VOR. These outlets have receive-only capability and rely on a VOR or a remote transmitter for full capability. See also REMOTE COMMUNICATIONS OUTLET.

limited route concept
A concept of controlled airspace organisation which requires an aircraft operator to choose between a limited number of specified ATS routes for a flight from one point to another.

limit loads (aircraft)
The maximum loads assumed to occur in the anticipated operating conditions.

linearity sector
A sector containing the course line or ILS glide path, within a course sector or an ILS glide path sector, respectively, in which the increment of DDM per unit of displacement remains substantially constant.

line of position
In general, any line along which certain navigational values are constant, as a line of constant bearing towards a given radio station.

liquid-crystal display
Digital display made by changing the reflectivity of 'liquid crystals' (alignments of molecules) by an applied electric field.

lithometeor
The general term for dry particles suspended in the atmosphere such as dust, haze, smoke and sand.

litre
The volume of 1 kg water at 4°C; a cubic decimetre; 1000 cc or $1/1000\,m^3$.

load factor
The ratio of a specified load to the total weight of the aircraft. The specified load is expressed in terms of any of the following: aerodynamic forces, inertia forces, or ground or water reactions.

load threshold
A value, expressed in movements per hour, used in conjunction with air traffic counts, to generate warnings. This value should represent the level of plannable traffic, i.e. traffic which could be expected to be known to the central data bank, above which capacity problems are likely to be encountered.

localiser
The component of an ILS that provides course guidance to the runway. See also INSTRUMENT LANDING SYSTEMS.

localiser course bend
A course bend is an aberration of the localiser course line with respect to its nominal position.

localiser course ILS [ICAO]
The locus of points, in any given horizontal point, at which the DDM (difference in depth of modulation) is zero.

localiser type directional aid [US]
A navaid, used for non-precision instrument approaches, with utility and accuracy comparable to a localiser, but which is not a part of a complete ILS and is not aligned with the runway.

localiser usable distance [US]
The maximum distance from the localiser transmitter at a specified altitude, as verified by flight inspection, at which reliable course information is continuously received.

local oscillator
Produces the radio frequency oscillation with which the received wave is combined.

local traffic
Aircraft operating in the traffic pattern or within sight of the tower, or aircraft known to be departing or arriving from flight in local practice areas, or aircraft executing practice instrument approaches at the airport. See also TRAFFIC PATTERN.

location indicator
A four-letter code group formulated in accordance with rules prescribed by ICAO and assigned to the location of an aeronautical fixed station.

locator
An LF/MF NDB used as an aid to final approach.
Note: A locator usually has an average radius of rated coverage of between 18.5 and 46.3 km (10 and 25 NM).

log book
In the case of an aircraft log book, engine log book or variable pitch propeller log book, or personal flying log book, includes a record kept either in a book, or by any other means approved by the Authority in the particular case.

longitude
Measured in degrees east and west of Greenwich from imaginary lines joining the poles called meridians, which are half great circles.

longitudinal axis (aeroplane)
A selected axis parallel to the direction of flight at a normal cruising speed, and passing through the centre of gravity of the aeroplane.

longitudinal separation
The longitudinal spacing of aircraft at the same altitude by a minimum distance expressed in units of time or miles. See also SEPARATION.

longitudinal waves
Waves in which the particles are displaced in the same direction as the wave: compression waves or sound waves.

Loran
A system of radio navigation in which position is determined by the relative travel time of pulse signals from two or more radio stations (a 'master' and one or more 'slave' stations). Standard Loran ('Loran A') is useful for distances up to 900 nautical miles by day and 1400 nautical miles by night; originally introduced in the 1940s, this pulse hyperbolic system operated on frequencies in the 1.9 MHz band, now obsolete. Loran C, for even greater distances, is a replacement system for Loran A and uses pulse hyperbolic techniques, operating on 100 kHz. Loran D is a shorter range version of Loran C.

Loran pair
A synchronised master station and slave station operating on the same radio frequency with the same pulse repetition period and serving a particular geographic area.

Loschmidt's constant
The number of molecules per cc of a perfect gas at standard temperature and pressure $= 2.68719 \times 10^{19}$.

lost communications
Loss of the ability to communicate by radio. Standard pilot procedures are specified in Aeronautical Information Publications (AIP). Radar controllers issue procedures for pilots to follow in the event of lost communications.

low
An area of low atmospheric pressure; a depression or cyclone.

low altitude airway structure [US]
The network of airways serving aircraft operations up to but not including 18 000 ft amsl. See also AIRWAY.

low altitude alert
Check your altitude immediately. See also SAFETY ADVISORY.

low approach
An approach over an airport or runway following an instrument approach or a VFR approach including the go-around manoeuvre where the pilot intentionally does not make contact with the runway.

low DDM zone
A zone outside a course sector or an ILS glide path sector in which the DDM is less than the minimum value specified for the zone.

lower sideband
The sideband of an AM transmission which is of lower frequency than the carrier.

low frequency
The radio frequency band between 30 and 300 kHz.

low modulation rates
Modulation rates up to and including 300 bauds.

lumen (lm)
The luminous flux emitted in a solid angle of 1 steradian by a point source having a uniform intensity of 1 candela.

luminescence
The emission of light from any cause except high temperature, e.g. the fluorescence and phosphorescence.

luminous paint
Paint made of phosphorescent substances (calcium sulphide) which glow in the dark. Used on watch dials and hands.

lux (lx)
The illuminance produced by a luminous flux of 1 lumen uniformly distributed over a surface of $1 \, m^2$.

machmeter
A temperature-linked air speed indicator which indicates the Mach number at which the aeroplane is flying.

mach number
The speed of the aircraft (TAS) divided by the speed of sound at that temperature.

magnesium
Atomic number 12, relative density 1.74, melting point 651°C. Burns with a brilliant white flame (photography). Used to make lightweight alloys.

magnet
(1) Permanent. Any ferromagnetic substance with a magnetic field which is independent of an electric current.
(2) Electro. Formed by passing a current throught a coil wound around a core (usually soft iron).

magnetic compass
An instrument indicating magnetic directions by means of a freely suspended compass card; the primary means of indicating the heading or direction of flight of an aircraft.

magnetic declination
Engineering term of magnetic variation.

magnetic dip
The angle of inclination of the earth's magnetic field below the horizontal.

magnetic equator
The aclinic line of zero magnetic dip which lies to the south of the equator in America and the eastern Pacific, and to the north of the equator across the Indian Ocean and Africa.

magnetic flux
A phenomenon produced in the medium surrounding electric currents or magnets. The amount of flux through any area is measured by the quantity of electricity caused by flow in a circuit of given resistance bounding the area when this circuit is removed from the magnetic field.

magnetic storms
Disturbances in the earth's magnetic field which cause compass and radio irregularities.

magnetic variation
The angular difference between true north and magnetic north.
Note: The value given indicates whether the angular difference is east or west of true north.

magnetism (terrestrial) or geomagnetism
The cause of the earth's magnetic field is not known. It has decreased by 5% in the last 100 years. It is defined by the three 'magnetic elements': the horizontal component (of the flux density); the angle of dip (of the flux density); and the angle of variation (declination) from the axis.

magnetometer
An instrument for comparing the strengths of magnetic fields.

magnetosphere
The space occupied by the earth's magnetic field.

magnetron
A thermionic valve used in radar to produce oscillations in the microwave region.

mail
Dispatches of correspondence and other objects tendered by, and intended for delivery to postal administrations.

main rotor
The rotor that supplies the principal lift to a rotorcraft.

main runway
The runway most used for takeoff and landing.

maintain
Concerning altitude/flight level, to remain at the altitude/flight level specified. The phrase 'climb and' or 'descend and' normally preceeds 'maintain' and the altitude assignment: e.g. 'descend and maintain 5000 ft'. If a SID procedure is assigned in the initial or subsequent clearance, the altitude restrictions in the SID, if any, will apply unless otherwise advised by ATC.

maintenance
Inspection, overhaul, repair, preservation and replacement of parts, but excluding preventive maintenance.

Maintenance Review Board [UK]
A Board set up as part of the Type Approval procedures, to establish a maintenance programme for aircraft exceeding 5700 kg MTWA intended for certification in the Transport Category.

Maintenance Schedule
Maintenance performed at defined intervals to specific items of aircraft equipment as defined in the Maintenance Schedule, including Scheduled Maintenance Inspections, replacements, adjustments, calibration and cleaning.

Maintenance Statement [UK]
A statement issued with the Certificate of Maintenance Review as an entry into the Technical Log, detailing when the next Scheduled Maintenance Inspection and/or any out of phase inspection is due.

major alteration
An alteration not listed in the aircraft, aircraft engine or propeller specifications:

(1) That might appreciably affect weight, balance, structural strength, performance, powerplant operation, flight characteristics or other qualities affecting airworthiness; or
(2) That is not done according to accepted practices or cannot be done by elementary operations.

major repair
A repair:

(1) That, if improperly done, might appreciably affect weight, balance, structural strength, performance, powerplant operation, flight characteristics or other qualities affecting airworthiness; or
(2) That is not done according to accepted practices or cannot be done by elementary operations.

make short approach
Used by ATC to inform a pilot to alter his traffic pattern so as to make a short final approach.

making way
An aeroplane on the surface of the water is 'making way' when it is under way and has a velocity relative to the water.

Mandatory Aircraft Modifications and Inspection Summary [UK]
A CAA publication that summarises the mandatory modifications and inspections to be complied with for aircraft and equipment manufactured and registered in the United Kingdom.

mandatory altitude
An altitude depicted on an instrument approach procedure chart requiring the aircraft to maintain altitude at the depicted value.

manganese
Atomic number 25, relative density 7.2, melting point 1244°C. Used in many alloys.

manifold pressure
Absolute pressure as measured at the appropriate point in the induction system and usually expressed in inches of mercury.

manoeuvring area
That part of an aerodrome provided for the takeoff and landing of aircraft and for the movement of aircraft on the surface, excluding the apron and any part of the aerodrome provided for the maintenance of aircraft.

manometer
An instrument for measuring gas pressure.

Mares' tails
A pattern of cirrus clouds.

marker
An object displayed above ground level in order to indicate an obstacle or delineate a boundary.

marker beacon
An electronic navigation facility transmitting a 75 MHz vertical fan or bone-shaped radiation pattern. Marker beacons are identified by their modulation frequency and keying code and, when received by compatible airborne equipment, indicate to the pilot, both aurally and visually, that he is passing over the facility. See also INNER MARKER, MIDDLE MARKER and OUTER MARKER.

marking
A symbol or group of symbols displayed on the surface of the movement area in order to convey aeronautical information.

maser
Microwave amplification by stimulated emission of radiation, using the internal energy of atoms to produce 'low-noise' microwave amplifiers and oscillators.

maximum and minimum thermometer
A thermometer which uses both alcohol and mercury to record the highest and lowest temperatures each day. Also called Six's thermometer.

maximum authorised altitude [US]
A published altitude representing the maximum usable altitude or flight level for an airspace structure or route segment. It is the highest altitude on a federal airway, jet route, area navigation low or high route, or other direct route for which an MEA is designated in FAR Part 95, at which adequate reception of navigation and signals is assured.

maximum mass
Maximum certificated takeoff mass.

maximum total weight authorised [UK]
The maximum total weight of the aircraft and its contents at whichs the aircraft may take off anywhere in the world, in the most favourable circumstances and in accordance with the Certificate of Airworthiness.

maximum wind axis
On a constant pressure chart, a line denoting the axis of maximum wind speeds at that constant pressure surface.

mayday
The international radiotelephony distress signal, which is repeated three times. It indicates imminent and grave danger and that immediate assistance is requested.

mean
Average of a set of values.

mean power (of a radio transmitter)
The average power supplied to the antenna transmission line by a transmitter during an interval of time sufficiently long compared with the lowest frequency encountered in the modulation taken under normal operating conditions.
Note: A time of 1/10 second during which the mean power is greatest will normally be selected.

mean sea level
The average height of the surface of the sea for all stages of tide, used as reference for elevations.

mean time
Time measured by the rotation of the earth with respect to the mean sun.

mean time between failures
The actual operating time of a facility divided by the total number of failures of the facility during that period of time.
Note: The operating time should in general be chosen so as to include at least five, and preferably more, facility failures in order to give a reasonable measure of confidence in the figure derived.

measured ceiling
A ceiling classification applied when the ceiling value has been determined by instruments or the known heights of unobscured portions of objects other than natural landmarks.

mechanical advantage
The load divided by the force used to move it.

mechanical equivalent of heat
1 Joule = 0.238846 calories = 2.7777×10^{-7} kWh = 9.47813 BTU; 1 Joule = 1 newton metre = 0.737561 ft lb = 10^7 ergs = 1 watt second.

medical assessment
The evidence issued by an ICAO contracting State that the licence holder meets specific requirements of medical fitness. It is issued following an evaluation by the Licensing Authority of the report submitted by the designated medical examiner who conducted the examination of the applicant for the licence.

medical certificate
Acceptable evidence of physical fitness on a form prescribed by the aviation administration.

medium aircraft [ICAO]
For the purpose of weight turbulence categorisation, aircraft less than 136 000 kg and more than 7000 kg maximum take-off weight, are categorised as 'medium'.

medium frequencies
Radio frequencies in the 300–3000 kilohertz range.

megahertz (MHz)
One million hertz, i.e. one million cycles per second.

melting point
The temperature at which a substance may be either solid or liquid (at a pressure of one atmosphere, i.e. 760 mm mercury).

memory
A device which stores information on a computer or similar system for future use.

mensuration
The measurement of lengths, areas and volumes.

mercator
The chart projection commonly used for nautical charts. On this projection the rhumb line is represented by a straight line; greater circles (radio bearings are great circles) are represented by curved lines. Due to the rapidly expanding scale, distances must be measured with the scale for the middle latitude between the two points in question. It is also used for a number of long range navigation charts of the armed forces.

mercatorial bearing
A radio bearing converted from a great-circle direction to a rhumb-line direction (mercatorial bearing) for a plotting on a mercator chart.

mercurial barometer
A barometer in which pressure is determined by balancing air pressure against the weight of a column of mercury in an evacuated glass tube.

mercury (Hg)
Atomic number 80, relative density 13.6, melting point $-39°C$, boiling point 357°C. Used in thermometers, barometers and manometers.

meridians of longitude
360 imaginary lines (half great circles) joining the north and south poles. Measured 1° to 179° east and 1° to 179° west of the prime meridian 0° (Greenwich). The 180° meridian is common to both east and west and it is upon this meridian that the International Date Line is based.

message
A communication sent from one location to another and comprising an integral number of fields.

message element
The smallest assembly of characters, in a message, which has an independent meaning.   *Note:* A message element is analogous to a word in plain language.

message field
An assigned area of a message containing specified elements of data.

message format
The disposition and structure of the message fields which constitute a message.

metabolism
The storage of energy by the breaking down of complex substances (catabolism).

metal fatigue   See FATIGUE OF METAL.

metastable state
The state of super-cooled water or super-saturated solutions. The change of state will occur only if a crystal is added.

meteorological authority
The authority providing or arranging for the provision of meteorological service for international air navigation on behalf of a contracting State.

meteorological bulletin
A text comprising meteorological information preceded by an appropriate heading.

meteorological information
Meteorological report, analysis, forecast and any other statement relating to existing or expected meteorological conditions.

meteorological office
An office designated to provide meteorological service for international air navigation.

meteorological operational channel
A channel of the aeronautical fixed service (AFS), for the exchange of aeronautical meteorological information.

meteorological operational telecommunications network
An integrated system of meteorological operational channels, as part of the aeronautical fixed service, for the exchange of aeronautical meteorological information between the aeronautical fixed stations within the network.
Note: 'Integrated' is to be interpreted as a mode of operation necessary to ensure that the information can be transmitted and received by the stations within the network in accordance with pre-established schedules.

meteorological report
A statement of observed meteorological conditions related to a specified time and location.

meteorological satellite
An artificial earth satellite making meteorological observations and transmitting these observations to earth.

meteorological visibility [US]
In observing practice, a main category of visibility which includes the sub-categories of prevailing visibility and runway visibility. Meteorological visibility is a measure of horizontal visibility near the earth's surface, based on sighting of objects in the daytime or unfocused lights of moderate intensity in the daytime, or unfocused lights of moderate intensity at night. Compare slant visibility, runway visual range, vertical visibility. See also SECTOR VISIBILITY, SURFACE VISIBILITY and TOWER VISIBILITY.

meteorology
The scientific study of the weather.

Meteosat
The European geostationary meteorological satellite.

metering [US]
A method of time regulating arrival traffic flow into a terminal area so as not to exceed a predetermined terminal acceptance rate.

metering fix [US]
A fix along an established route over which aircraft will be metered prior to entering terminal airspace. Normally this fix should be established at a distance from the airport that will facilitate a profile descent 10 000 ft above airport elevation or above.

metre
The distance travelled by light in a vacuum during 1/2 99 792 458 of a second.

metrology
The scientific study of weights and measures.

microbarograph
An instrument which records very small and rapid variations of atmospheric pressure.

microcomputer
A complete digital computing system the hardware of which comprises a microprocessor and other large scale integration circuits such as memory and input/output ports; a single chip with circuits capable of providing control, arithmetic/logic operations, memory and input/output.

microelectronics
Electronics using minute solid state components largely based on integrated circuits (silicon chips).

microlight [UK]
An aeroplane with a maximum total weight authorised not exceeding 390 kg, a wing loading at the maximum total weight authorised not exceeding 25 kg per square metre, a maximum fuel capacity not exceeding 50 litres and which has been designed to carry not more than two persons.

micrometer
Device for measuring very small distances and angles.

micron
One thousandth of a millimetre.

microphone
An instrument for converting sound waves into electrical impulses. In consists of a diaphragm which by vibrating alters:

(a) the capacitance of a condenser; or
(b) the EMF of a piezoelectric crystal; or
(c) the conductivity of packed carbon granules.

microprocessor
A chip which provides the control and arithmetic/logic operations required by a digital computer; a chip capable of processing information in digital form in accordance with a coded and stored set of instructions in order to control the operation of other circuitry or equipment.

microstrip
Transmission lines and passive components formed by depositing metal strips of suitable shapes and dimensions on one side of a dielectric substrate, the other side of which is completely coated with metal acting as a ground plane.

microwave landing system
An instrument landing system operating in the microwave spectrum which provides lateral and vertical guidance to aircraft having compatible avionics equipment. See also INSTRUMENT LANDING SYSTEMS.

microwaves
A vague term used to describe radio frequencies above 1000 MHz.

middle compass locator (LMM)
A compass locator installed in conjunction with the middle marker of an instrument landing system. See also MIDDLE MARKER.

middle marker
A marker beacon that defines a point along the glide slope of an ILS normally located at or near the point of decision height (ILS category I). See also FAN MARKER and INSTRUMENT LANDING SYSTEM.

mil
One thousandth of an inch.

mile, nautical
The ordinary unit of 6,076.1 feet (1,852.0 metres) for measuring distances at sea, and now generally used in air navigation. For practical purposes a minute of latitude may be considered as equal to a nautical mile. It is approximately equal to 1.15 statute miles.

military aircraft [UK]
'Military aircraft' include the naval, military or air force aircraft of any country and:

(1) any aircraft being constructed for the naval, military or air force of any country under a contract entered into by the Secretary of State; and
(2) any aircraft in respect of which there is in force a certificate issued by the Secretary of State that the aircraft is to be treated for the purposes of this Order as a military aircraft.

military training area
An area of airspace of defined dimensions within which intense military flying training takes place. Because of the random nature of the activity within these areas it is not always possible to provide civil air traffic control service during the published hours of activity.

military training routes [US]
Airspace of defined vertical and lateral dimensions established for the conduct of military flight training at airspeeds in excess of 250 knots IAS. See also IFR MILITARY TRAINING ROUTES.

millibar
Unit of atmospheric pressure. 1000 dynes per cm^2, 100 newtons per m^2, 1000 millibars being equal to 29.53 inches of mercury; conversely, 1 inch of mercury is equal to 33.86 millibars.

mineral oil   See PETROLEUM.

minima
Weather condition requirements established for a particular operation or type of operation, e.g. IFR takeoff or landing, alternate airport for IFR flight plans, VFR flight. See also DEPARTURE PROCEDURES and LANDING MINIMA.

minimum descent altitude
The lowest altitude, expressed in feet above mean sea level, to which descent is authorised on final approach or during circle-to-land manoeuvring in execution of a standard instrument approach procedure, where no electronic glideslope is provided, i.e. non-precision approach.

minimum descent height
In relation to the operation of an aircraft at an aerodrome, the height in a non-precision approach below which descent may not be made without the required visual reference.

minimum en-route IFR altitude [US]
The lowest published altitude between radio fixes that assures acceptable navigational signal coverage and meets obstacle clearance requirements between those fixes. The MEA prescribed for a Federal airway or segment thereof, area navigation low or high route or other direct route, applies to the entire width of the airway, segment or route between the radio fixes defining the airway, segment or route.

minimum fuel
Indicates that an aircraft's fuel supply has reached a state where, upon reaching the destination, it can accept little or no delay. This is not an emergency situation but merely indicates an emergency situation is possible should any undue delay occur.

minimum holding altitude
The lowest altitude prescribed for a holding pattern that assures navigational signal coverage and communications, and meets obstacle clearance requirements.

minimum glide path
The lowest angle of descent along the zero degree azimuth that is consistent with published approach procedures and obstacle clearance criteria.

minimum IFR altitude [US]
Minimum altitudes for IFR operations as prescribed in FAR Part 91. These altitudes are published on aeronautical charts and prescribed in FAR Part 95, for airways and routes, and FAR Part 97 for standard instrument approach procedures. If no

applicable minimum altitude is prescribed in FAR Parts 95 or 97, the following minimum IFR altitude applies:

(1) In designated mountainous areas, 2000 ft above the highest obstacle within a horizontal distance of five statute miles from the course to be flown; or

(2) Other than mountainous areas, 1000 ft above the highest obstacle within a horizontal distance of five statute miles from the course to be flown; or

(3) As otherwise authorised by the administrator or assigned by ATC. See also MINIMUM EN-ROUTE IFR ALTITUDE, MINIMUM OBSTRUCTION CLEARANCE ALTITUDE, MINIMUM SAFE ALTITUDE and MINIMUM VECTORING ALTITUDE.

minimum obstruction clearance altitude [US]

The lowest published altitude in effect between radio fixes on VOR airways, off-airway routes, or route segments which meets obstacle clearance requirements for the entire route segment and which assures acceptable navigational signal coverage only within 22 NM of the VOR.

minimum off-route altitude [US]

An altitude derived by Jeppesen. The MORA provides terrain and obstruction clearance within 10 NM of the route centreline (regardless of the route width) and end fixes. A grid MORA altitude provides terrain and obstruction clearance within the section outlined by latitude and longitude lines. A MORA at 7000 ft or less clears all obstructions and terrain by 1000 ft; a MORA greater than 7000 ft clears all terrain by 2000 ft. When a MORA is shown along a route as 'unknown' or within a grid as 'unsurveyed' a MORA altitude is not shown due to incomplete or insufficient official source.

minimum reception altitude

The lowest altitude at which an intersection can be determined.

minimum safe altitude

(1) The minimum altitude at which to fly for various aircraft operations.

(2) Altitudes depicted on approach charts that provide at least 1000 ft of obstacle clearance for emergency use within a specified distance from the navigation facility upon which a procedure is predicated. These altitudes will be identified as minimum sector altitudes or emergency safe altitudes and are established as follows:

(a) Minimum sector altitudes. Altitudes depicted on approach charts that provide at least 1000 ft of obstacle clearance within a 25 mile radius of the navigation facility upon which the procedure is predicated. Sectors depicted on approach charts must be at least 90° in scope. These altitudes are for emergency use only and do not necessarily assure acceptable navigational signal coverage.

(b) Minimum sector altitude (ICAO). The lowest altitude that may be used under emergency conditions that will provide a minimum clearance of 300 m (1000 ft) above all obstacles located in an area contained within a sector of a circle of a 25 nm radius centred on a radio aid to navigation.

(c) Emergency safe altitudes. Altitudes depicted on approach charts that provide at least 1000 ft of obstacle clearance in non-mountainous areas and 2000 ft of obstacle clearance in designated mountainous areas within a 100 NM radius of the navigation facility upon which the procedure is predicated and normally used only in military procedures. These altitudes are identified on published procedures as 'emergency safe altitudes'.

minimum safe altitude warning [US]
A function of the ARTS III computer that aids the controller by alerting him when a tracked Mode C equipped aircraft is below or is predicted by the computer to go below a predetermined minimum safe altitude.

minimums
Alternate plural for minima

minimum sector altitude
The lowest altitude which may be used under emergency conditions which will provide a minimum clearance of 300 m (984 ft) above all objects located in an area contained within a sector of a circle of 46 km (25 NM) radius centred on a radio aid to navigation.

minimum vectoring altitude
The lowest MSL altitude at which an IFR aircraft will be vectored by a radar controller, except as otherwise authorised for radar approaches, departures and missed approaches. The altitude meets IFR obstacle clearance criteria. It may be utilised for radar vectoring only upon the controller's determination that an adequate radar return is being received from the aircraft being controlled.

mishandled baggage
Baggage involuntarily or inadvertently separated from passengers or crew.

missed approach
(1) A manoeuvre conducted by a pilot when an instrument approach cannot be completed to a landing. The route of flight and altitude are shown on instrument approach procedure charts.
(2) A term used by the pilot to inform ATC that he is executing the missed approach.
(3) At locations where ATC radar service is provided, the pilot should conform to radar vector, when provided by ATC, in lieu of the published missed approach procedure. See also MISSED APPROACH POINT.

missed approach point
That point in an instrument approach procedure at or before which the prescribed missed approach procedure must be initiated in order to ensure that the minimum obstacle clearance is not infringed.

missed approach procedure [ICAO]
The procedure to be followed if, after an instrument approach, a landing is not effected and occurring normally:

(1) When the aircraft has descended to the decision height and has not established visual contact; or,

(2) When directed by air traffic control to go around again.

mixing ratio
The ratio by weight of the amount of water vapour in a volume of air to the amount of dry air; usually expressed as grams per kilogram (g/kg).

MLS approach reference datum
A point on the minimum glide path at a specified height above the threshold.

MLS back azimuth reference datum
A point at a specified height above the runway centreline at the runway midpoint.

MLS datum point
The point on the runway centreline closest to the phase centre of the approach elevation antenna.

mobile surface station
A station in the aeronautical telecommunications service, other than an aircraft station, intended to be used while in motion or during halts at unspecified points.

mode
The letter or number assigned to a specific pulse spacing of radio signals transmitted or received by ground interrogator or airborne transponder components of the air traffic control radar beacon system (ATCRBS). Mode A (military Mode 3) and Mode C (altitude reporting) are used in air traffic control. See also INTERROGATOR, RADAR and TRANSPONDER.

mode (SSR mode) [ICAO]
The letter or number assigned to a specific pulse spacing of the interrogation signals transmitted by an interrogator. There are 4 modes, A, B,C and D corresponding to four different interrogation pulse spacings.

mode W, X, Y, Z
A method of coding the DME transmissions by time spacing pulses of a pulse pair, so that each frequency can be used more than once.

modification
Modifications are changes made to a particular aircraft including: engines, propellers, radio apparatus, accessories, instruments, equipment and their installations. Substitution of one type for another, when applied to components, engines, propellers, radio apparatus, accessories, instruments and equipment is also considered to be a modification.

modulation rate
The reciprocal of the unit interval measured in seconds. This rate is expressed in bauds.
Note: Telegraph signals are characterised by intervals of time of duration equal to or longer than the shortest or unit interval. The modulation rate (formerly telegraph speed) is therefore expressed as the inverse of the value of this unit interval. If, for example, the unit interval is 20 milliseconds, the modulation rate is 50 bauds.

module
(1) A replaceable/detachable unit.
(2) A detachable section of a spacecraft.

moist-adiabatic lapse rate   See SATURATED-ADIABATIC LAPSE RATE.

moisture
An all-inclusive term denoting water in any or all of its three states.

mole (mol)
The amount of substance of a system which contains as many elementary entities as there are atoms in 0.012 kg of carbon mass number 12.
Note: When the mole is used, the elementary entities must be specified and may be atoms, molecules, ions, electrons, other particles or specified groups of such particles.

molybdenum
Atomic number 42, relative density 10.2, melting point 2620°C. Used for special steels and alloys.

moment of force
If the direction of a force is not towards the centre of gravity of the body, it tends to rotate the body. The moment of force is calculated by: force × perpendicular distance from the axis of rotation.

momentum
Mass × velocity, $kg \cdot ms^{-1}$

monocoque
The form of stressed skin used in a fuselage or nacelle in which the curved skin carries the whole or greater part of the main load.

monsoon
A wind that reverses its direction with the season, blowing more or less steadily from the interior of a continent towards the sea in winter, and in the opposite direction in the summer.

month
The 'calendar month' is defined by the Gregorian calendar; the 'solar month' is one twelfth of a solar year; the 'lunar month' is usually measured as a 'synodic month', i.e. 29.5306 days, being the period between two successive phases of the moon.

moon
Satellite of the earth. Its mass is 0.0123 that of the earth; its diameter is 3476 km, and its mean distance from the earth is approximately 382 000 km. Its escape velocity is only 2.4 km/s which is too low ever to hold an atmosphere of gas molecules. Its gravity is one sixth that of the earth.

mosaic
Several runs of vertical photographs either pinned over each other or cut out and pasted onto a backcloth. They can be re-photographed.

motion sickness
A clinical manifestation of vestibular excitation (it is known that individuals without a functioning vestibular apparatus do not suffer from this condition). Air sickness, like car sickness, or sea sickness, is a relatively common condition which affects children and adults who are said to be of a nervous disposition. This apprehension, if it exists, is really the fear of the thought of being sick and this fact

alone goes a long way to explain why individuals of the stomach and courage of Nelson can be incapacitated by motion sickness. Drugs recommended for the relief of motion sickness should be taken only on medical advice and should never be prescribed for the lone aviator, because motion sickness drugs produce cerebral sedation, as well as reducing vestibular response to rotational movements.

mountain waves
In certain weather conditions the disturbance of a transverse airflow by high ground can create an organised flow pattern of waves and large scale eddies in which strong up-draughts and down-draughts and turbulence frequently occur. These organised flow patterns are usually called mountain waves, but may also be referred to as lee waves or standing waves, and can be associated with relatively low hills and ridges as well as with high mountains. Mountain wave systems may extend for many miles down wind of the initiating high ground. Satellite photographs have shown wave clouds extending more than 250 NM from the Pennines and as much as 500 NM down wind of the Andes. However, 50 to 100 NM is a more usual extent of wave systems in most areas. Wave systems may, on occasion, extend well into the stratosphere.

movement area
The runways, taxiways and other areas of an airport that are utilised for taxi-ing, takeoff and landing of aircraft, exclusive of loading ramp and parking areas. At those airports with a tower, specific approval for entry onto the movement area must be obtained from ATC.

movement area [ICAO]
That part of an aerodrome intended for the surface movement of aircraft, including manoeuvring areas and aprons.

moving target indicator
An electronic device that will permit radar scope presentation only from targets that are in motion. A partial remedy for ground clutter.

multi-channel receiver
A global positioning satellite receiver that can simultaneously track more than one satellite signal.

multipath errors
Errors caused by the interference of a signal that has reached the receiver antenna by two or more different paths. Usually caused by one path being bounced or reflected.

multiplexing channel
A channel of a global positioning satellite receiver that can be sequenced through a number of satellite signals.

mutual inductance
The EMF induced in one circuit caused by a rapidly changing magnetic field in an adjacent circuit. SI unit: henry.

myopia

Short sight; inability to see distant objects. This is, that state where, when the eye is fully relaxed, parallel rays of light are focused in front of the retina either because the eyeball axis is too long or the refractive power of the eye is too strong. The condition is treated with concave lenses which should be worn at all times for distance vision.

N

nacelle
Structure usually streamlined to accommodate engines (or crew).

NAS–stage A [US]
The en-route ATC system's radar, computers and computer programs controller plan view displays (PVDs/radar scopes), input/output devices and the related communications equipment that are integrated to form the heart of the automated IFR air traffic control system. This equipment performs flight data processing (FDP) and radar data processing (RDP). It interfaces with automated terminal systems and is used in the control of en-route IFR aircraft.

National Airspace System
The common network of airspace; air navigation facilities, equipment and services, airports or landing areas; aeronautical charts, information and services; rules, regulations and procedures, technical information, manpower and material. Included are system components shared jointly with the military.

national beacon code allocation plan airspace [US]
Airspace over US territory located within the North American continent between Canada and Mexico, including adjacent territorial waters outward to abut boundaries of oceanic control areas/flight information areas.

national flight data centre
A facility in Washington DC, established by the FAA to operate a central aeronautical information service for the collection, validation and dissemination of aeronautical data in support of the activities of government, industry and the aviation community. The information is published in the *National Flight Data Digest*.

National Flight Data Digest [US]
A daily (except at weekends and on Federal holidays) publication of flight information appropriate to aeronautical charts, aeronautical publications, Notices to Airmen or other media serving the purpose of providing operational flight data essential to safe and efficient aircraft operations.

National Search and Rescue Plan
An interagency agreement that provides for the effectives utilisation of all available facilities in all types of search and rescue missions.

nautical mile
Defined in the UK as 6080 ft; international NM = 1852 m.

navaid
Any type of navigation aid, especially electronic aids which are sited at a ground station, latterly including satellite systems.

navaid classes
VOR, VORTAC and TACAN aids are classed according to their operational use. These three classes of ground based navaids are:

(1) T. Terminal.
(2) L. Low altitude.
(3) H. High altitude.

navigable airspace
Airspace at and above the minimum flight altitudes prescribed by national legislation including airspace needed for safe takeoffs and landings.

navigation
The process of directing an aircraft so as to reach an intended destination. Types of navigation are:

(1) Celestial. The determination of position by means of sextant observations of the celestial bodies, together with the exact time of observations.
(2) Dead reckoning. The determination of the distance and direction between two known points, or the determination of position from a knowledge of the distance and direction from a known point.
(3) Electronic. Navigation and determination of position by electronic means.
(4) Pilotage. Directing an aircraft with respect to visible landmarks; sometimes broadened to include landmarks recognisable by optical, aural, mechanical or electronic means.
(5) Radio. The determination of position by means of observed radio signals, from ground based installations or satellites.

navigational aid
Any visual or electronic device airborne or on the surface that provides point-to-point guidance information or position data to aircraft in flight. See also AIR NAVIGATION FACILITY.

negative
'No'; 'Permission not granted'; 'That is not correct'.

negative contact
Used by pilots to inform ATC that:

(1) Previously issued traffic is not in sight. It may be followed by the pilot's request for the controller to provide assistance in avoiding the traffic.
(2) They were unable to contact ATC on a particular frequency.

negative pole
(1) The south-seeking pole of a magnet.
(2) The cathode of a battery.

neon (N)
Atomic number 10. A colourless inert gas which glows red when an electric current passes through it. Used to make coloured lights.

neoprene
A plastic material which has a high tensile strength and a better heat and ozone resistance than rubber.

nephanalysis
The graphical depiction of analysed cloud data on a geographical map.

net gradient
The net gradient of climb is the expected gradient of climb diminished by the manoeuvre performance (i.e. that gradient of climb necessary to provide power to manoeuvre) and by the margin (i.e that gradient of climb necessary to provide for those variations in performance which are not expected to be taken explicit account of operationally).

net safety benefit analysis
A method of analysis used to determine the safety benefits of systems or procedures relative to aircraft operations.

network station
An aeronautical station forming part of a radiotelephony network.

New African Air Transport Policy
Conference of African ministers in charge of civil aviation.

newton (N)
The force which, when applied to a body having a mass of 1 kilogram, gives it acceleration of $1\,m/s^2$.

Newtonian mechanics
Developed from Newton's laws of motion, for the study of the motion of bodies at ordinary velocities.

nickel (Ni)
Atomic number 28, relative density 8.9, melting point 1455°C. Used for plating, steel, alloys and coinage.

nicotine
A colourless poisonous liquid found in tobacco.

night [ICAO]
The hours between the end of evening civil twilight and the beginning of morning civil twilight or such other period between sunset and sunrise as may be specified by the appropriate authority.

night effect
Displacement of a radio bearing due to interference from reflected sky waves or other causes peculiar to night conditions.

nimbostratus
A principal cloud type, grey coloured, often dark, the appearance of which is rendered diffuse by more or less continuously falling rain or snow, which is most cases reaches the ground. It is thick enough throughout to blot out the sun.

nit
A unit of luminance equal to one candela per square metre.

noctilucent clouds
Clouds of unknown composition which occur at great heights, probably around 75 to 90 km. They resemble thin cirrus, but usually have a bluish or silverish colour (sometimes orange to red), standing out against a dark night sky. Rarely observed.

nocturnal
Occurring during the hours between sunset and sunrise.

no-gyro approach/vector
A radar approach/vector provided in case of a malfunctioning gyrocompass or directional gyro. Instead of providing the pilot with headings to be flown, the controller observes the radar track and issues control instructions, 'turn right/left' or 'stop turn' as appropriate.

noise (radio)
Occurs in amplifying circuits due to voltage variations caused by random movements of electrons, vibration of components etc.; a disturbance which is not part of the message.

noise abatement procedure
A procedure which requires operators of aircraft to limit or mitigate the effect of noise or vibration during takeoff or landing at specified aerodromes.

non-approach control tower [US]
Authorises aircraft to land or takeoff at the airport controlled by the tower, or to transit the airport traffic area. The primary function of non-approach control tower is the sequencing of aircraft in the traffic pattern and on the landing area. Non-approach control towers also separate aircraft operating under instrument flight rules clearances from approach controls and centres. They provide ground control services to aircraft, vehicles, personnel and equipment on the airport movement area.

non-composite separation [US]
Separation in accordance with minima other than the composite separation minimum specified for the area concerned.

non-directional beacon
An L/MF or UHF radio beacon transmitting non-directional signals whereby the pilot of an aircraft equipped with direction finding equipment can determine his bearing to or from the radio beacon and 'home' on or track to or from the station. When the radio beacon is installed in conjunction with the instrument landing system marker, it is normally called a compass locator. See also AUTOMATIC DIRECTION FINDER and COMPASS LOCATOR.

non-directional radio beacon   See NON-DIRECTIONAL BEACON.

non-ferrous metal
Any metal or alloy which does not contain iron.

non-instrument runway
A runway intended for the operation of aircraft using visual approach procedures.

non-network communications
Radiotelephony communications conducted by a station of the aeronautical mobile service, other than those conducted as part of a radiotelephony network.

non-precision approach procedure
A standard instrument approach procedure in which no electronic glideslope is provided, e.g. VOR, TACAN, NDB, LOC, ASR, LDA or SDF approaches.

non-radar
Precedes other terms and generally means without the use of radar, such as:

(1) Non-radar route. A flight path or route over which the pilot is performing his own navigation. The pilot may be receiving radar separation, radar monitoring or other ATC services while on a non-radar route. See also RADAR ROUTE.
(2) Non-radar approach. Used to describe instrument approaches for which course guidance on final approach is not provided by ground-based precision or surveillance radar. Radar vectors to the final approach course may or may not be provided by ATC. Examples of non-radar approaches are VOR, ADF, TACAN, and ILS approaches. See also FINAL APPROACH COURSE, FINAL APPROACH IFR, INSTRUMENT APPROACH PROCEDURE and RADAR APPROACH.
(3) Non-radar separation. The spacing of aircraft in accordance with established minima without the use of radar, e.g. vertical, lateral or longitudinal separation. See also RADAR SEPARATION.
(4) Non-radar separation (ICAO). The separation used when aircraft position information is derived from sources other than radar.
(5) Non-radar arrival. An arriving aircraft that is not being vectored to the final approach course for an instrument approach or towards the airport for a visual approach. The aircraft may or may not be in a radar environment and may or may not be receiving radar separation, radar monitoring or other services provided by ATC. See also RADAR ARRIVAL.
(6) Non-radar approach control. An ATC facility providing approach control service without the use of radar. See also APPROACH CONTROL and APPROACH CONTROL SERVICE.

no procedure turn
No procedure turn is required nor authorised without ATC clearance.

normal (meteorology)
The average value of a meteorological element over any fixed period of years that is recognised as standard for the country and for the element of concern.

normalising
Heat treatment applied to steel to remove internal stresses.

North American route
A numerically coded route pre-planned over existing airway and route systems to and from specific coastal fixes serving the North Atlantic. North American routes consist of the following:

(1) Common route/portion. That segment of a North American route between the inland navigation facility and the coastal fix.
(2) Non-common route/portion. That segment of a North American route between the inland navigation facility and a designated North American terminal.
(3) Inland navigation facility. A navigation aid on a North American route at which the common route and/or the non-common route begins or ends.
(4) Coastal fix. A navigation aid or intersection where an aircraft transitions between the domestic and oceanic route structures.

northerly turning error
Erratic behaviour in the indicated reading of an aircraft compass, resulting from turns when the aircraft is on a northerly (or southerly) heading.

NOTAM
A notice containing information concerning the establishment condition or change in any aeronautical facility, service, procedure or hazard, the timely knowledge of which is essential to personnel concerned with flight operations.

(1) Class I distribution. Distribution by means of telecommunications.
(2) Class II distribution. Distribution by means other than telecommunications.

no tail rotor anti-torque
A concept to replace the helicopter's conventional exposed tail rotor.

Notified
UK legislation: defined as 'set forth in a document published by the Civil Aviation Authority and entitled UK NOTAM or UK Aeronautical Information Publication'. References to the legislation under which notification is made are given in the sections of the UK Aeronautical Information Publication (*UK Air Pilot*).

no transgression zone
A concept used when simultaneous approaches on parallel or near parellel instrument runways occur. The zone is a corridor of airspace located centrally between the two extended runway centrelines. Its significance is that controllers must intervene if any of the aircraft is observed to penetrate the no-transgression zone (NTZ). The width of the NTZ depends on different airspace factors which need allowances such as:

(1) The limitation of the surveillance system and the controller reaction time in detecting the deviating aircraft as it enters the NTZ.
(2) The total time required for the controller to react, determine the approach resolution manoeuvre and communicate the appropriate command to the aircraft concerned, the time required for the pilot to understand the communication and react, and finally, for the beginning of aircraft response.
(3) Airspace needed for the completion of the resolution manoeuvre by the threatened aircraft.
(4) Adequate track separation in the lateral dimension, plus an allowance for the fact that the threatened aircraft may not be exactly on the extended runway centreline of the adjacent runway.

no-wind position
The position at which an aircraft would be at any given time, as a result of the distance and direction of its flight, if there had been no wind.

N-type conductivity
Caused by a flow of electrons in a semiconductor. (P-type is caused by a flow of holes.)

nuclear energy
The heat released when mass is converted into energy. Also called atomic energy.

numerical forecasting   See NUMERICAL WEATHER PREDICTION.

numerical weather prediction
Forecasting by digital computers solving mathematical equations. Used extensively in weather services throughout the world.

numerous targets vicinity (location)
A traffic advisory issued by ATC to advise pilots that targets on the radar scope are too numerous to issue individually.

nutation
The movement of the earth's poles around a mean position.

O

oblique line overlap
A run of oblique photographs normally overlapping by 60% in the foreground line.

obscuration
Sky hidden by surface-based obscuring phenomena and vertical visibility restricted overhead.

obscuring phenomena
Any hydrometeor or lithometeor other than clouds. May be surface based or aloft.

observation (meteorological)
The evaluation of one or more meterological elements.

obstacle
An existing object, object of natural growth or terrain at a fixed geographical location, or which may be expected at a fixed location within a prescribed area, with reference to which vertical clearance is or must be provided during flight operations.

obstacle assessment surface
A defined surface intended for the purpose of determining those obstacles to be considered in the calculation of obstacle clearance altitude/height for a specific ILS facility and procedure.

obstacle clearance altitude
The lowest altitude (OCA), or alternatively the lowest height above the elevation of the relevant runway threshold or above the aerodrome elevation as applicable (OCH), used in establishing compliance with appropriate obstacle clearance criteria.

obstacle clearance height   See OBSTACLE CLEARANCE ALTITUDE.

obstacle clearance limit
The height above aerodrome elevation below with the minimum prescribed vertical clearance cannot be maintained either on approach or in the event of a missed approach.

obstacle free zone
The airspace above the inner approach surface, inner transitional surfaces and balked landing surface, and that portion of the strip bounded by these surfaces which is not penetrated by any fixed obstacles other than a low-mass and frangibly mounted one required for air navigation purposes.

obstruction light
A light, or one of a group of lights, usually red or white, frequently mounted on a surface structure or natural terrain to warn pilots of the presence of an obstruction.

obtuse angle
An angle greater than 90°.

occluded front or occlusion
The front that is formed when and where a cold front overtakes a warm front or a stationary front.

oculogyral illusion
A visual illusion commonly encountered in night flying is known as the oculogyral illusion. If a source of light is viewed from an aircraft which itself is the subject of considerable angular acceleration, the light source will appear to develop a movement of its own and this apparent movement may continue for some time after straight and level flight has been resumed. This illusion is produced by involuntary eye movements (nystagmus) which are caused by vestibular inspired impulses reaching the brain from the semi-circular canals. The frequency and amplitude of these involuntary eye movements, and therefore the character of the illusion itself, will vary directly with the forces created by angular acceleration. It is for this reason that, while flying at night, care should be taken not to initiate manoeuvres which could create considerable angular acceleration forces.

off-load route(s)
Specific route(s) between given points or areas of origin and destination proposed by an air traffic flow management unit when the preferred route(s) is (are) not available or is (are) subject to delay.

off-route vector
A vector by ATC that takes an aircraft off a previously assigned route. Altitudes assigned by ATC during such vectors provide required obstacle clearance.

offset frequency simplex
A variation of single channel simplex wherein telecommunication betwen two stations is effected by using in each direction frequencies that are intentionally slightly different but contained within a portion of the spectrum allotted for the operation.

offset parallel runways
Staggered runways having centrelines that are parallel.

ohm (Ω)
Unit of resistance. Ohms = volts divided by amperes.

Omega
A long-range hyperbolic navigational system which operates on the very low frequency (VLF) band.

omni-bearing indicator
An instrument providing, by means of a left/right needle, continuous indication of the position of an aircraft relative to a selected omni-bearing.

omni-directional antenna
An antenna which radiates in or receives from all directions equally. Impossible to achieve in three dimensions.

omni-range
A VHF radio aid to navigation by means of which pilots may follow bearings to, or radials from, the facility in any desired direction. It also provides for off-course navigation.

omni-station
A VOR ground transmitter.

on course
(1) Used to indicate that an aircraft is established on the route centreline.
(2) Used by ATC to advise a pilot making a radar approach that his aircraft is lined up on the final approach course. See also ON-COURSE INDICATION.

on-course indication
An indication on an instrument that provides the pilot with a visual means of determining that the aircraft is located on the centreline of a given navigational track; or an indication on a radar scope that an aircraft is on a given track.

opaque
Does not allow waves (sound, light, X-rays) to pass through it; opposite of transparent or translucent.

open architecture
Applies to any system designed to accommodate attachment of additional, new features, thus reducing obsolescence.

operational control
The exercise of authority over the initiation, continuation, diversion or termination of a flight in the interests of the safety of the aircraft, and the regularity and efficiency of flight.

operational control communications
Communications required for the exercise of authority over the initiation, continuation, diversion or termination of a flight in the interests of the safety of the aircraft and the regularity and efficiency of a flight.
Note: Such communications are normally required for the exchange of messages between aircraft and aircraft operating agencies.

operational flight plan
The operator's plan for the safe conduct of the flight based on considerations of aeroplane performance, other operating limitations and relevant expected conditions on the route to be followed and at the aerodromes concerned.

operational performance ILS Category I [ICAO]
Operation down to 60 m (200 ft) decision height and with a runway visual range not less than a value of the order of 800 m (2600 ft) with a high probability of approach success.

operational performance ILS Category II [ICAO]
Operation down to 30 m (100 ft) decision height and with a runway visual range not less than a value of the order of 400 m (1200 ft) with a high probability of approach success.

operational performance ILS Category III A [ICAO]
Operation with no decision height limitation, to and along the surface of the runway with external visual reference during the final phase of the landing and with a runway visual range not less than a value of the order of 200 m (700 ft).

operational performance ILS Category III B [ICAO]
Operation, with no decision height limitation, to and along the surface of the runway without reliance on external visual reference and, sub-taxying in with external visual reference in a visibility corresponding to a runway visual range not less than a value of the order of 50 m (150 ft).

operational performance ILS Category III C [ICAO]
Operation, with no decision height limitation, to and along the surface of the runway and taxiways without reliance on external visual reference.

operational planning
The planning of flight operations by an operator.

operator
A person, organisation or enterprise engaged in, or offering to engage in, an aircraft operation.

optical maser
Laser.

option approach [US]
An approach requested and conducted by a pilot that will result in either a touch-and-go, missed approach, low approach, stop-and-go or full stop landing.

orbital velocity
The velocity required for a spacecraft to orbit round the earth (or moon). The synchronous orbital velocity of the earth is about 6260 knots.

organised track system
A moveable system of oceanic tracks that traverses the North Atlantic between Europe and North America, the physical positon of which is determined twice daily taking the best advantage of the winds aloft.

orientation
In radio navigation, a method of determining position relative to a radio station by means of a radio direction finder.

ornithopter
A heavier-than-air aircraft supported in flight chiefly by the reactions of the air on planes to which a flapping motion is imparted.

orographic
Of, pertaining to, or caused by mountains, as in orographic clouds, orographic lift or orographic precipitation.

orsat apparatus
For measuring the amounts of oxygen, carbon monoxide and carbon dioxide in exhaust gases.

orthomorphic projection
European term for the class of projections known as 'conformal'. Typical conformal (orthomorphic) projections are the Lambert, Mercator and stereographic.

oscillator
A transistor device with a resonant circuit for converting d.c. ir:to a.c.

oscilloscope
Uses a cathode ray tube to display images of varying electrical quantities. Used in radar systems.

osmium (Os)
Atomic number 76, relative density 22.57, melting point 3045°C. The heaviest substance. Used in alloys with platinum and iridium.

ounce
Weight = 28.3 g; volume = 28.41 cc; troy = 31.1 g.

out (RTF)
The conversation is ended and no response is expected.

outer compass locator (LOM)
A compass locator installed in conjunction with the outer marker of an instrument landing system. See also OUTER MARKER.

outer fix
A general term used within ATC to describe fixes in the terminal area, other than the final approach fix. Aircraft are normally cleared to these fixes by an air traffic controller. Aircraft are normally cleared from these fixes to the final approach fix, or final approach track.

outer marker
A marker beacon at or near the glideslope intercept altitude of an ILS approach. It is keyed to transmit two dashes per second on a 400 Hz tone that is received aurally and visually by compatible airborne equipment. The OM is normally located on the extended centreline of the runway. See INSTRUMENT LANDING SYSTEMS and MARKER BEACON.

out-of-coverage indication signal
A signal radiated into areas outside the intended coverage sector where required specifically to prevent invalid removal of an airborne warning indication in the presence of misleading guidance information.

over (RTF)
My transmission is ended; I expect a response.

overhaul
A major work operation which involves dismantling, bench testing and renewal of operational life.

overhead approach/360 overhead [US]
A series of predetermined manoeuvres prescribed for VFR arrival of military aircraft (often in formation) for entry into the VFR traffic pattern and to proceed to a landing. The pattern usually specifies the following:

(1) The radio contact required of the pilot.
(2) The speed to be maintained.
(3) An initial approach three to five miles in length.
(4) An elliptical pattern consisting of two 180° turns.
(5) A break point at which the first 180° turn is started.
(6) The direction of turns.
(7) Altitude (at least 500 ft above the conventional pattern).
(8) A 'roll-out' on final approach not less than one quarter mile from the landing threshold and not less than 300 ft above the ground.

over-the-top
Above the layer of clouds or other obscuring phenomena forming the ceiling.

oxides of nitrogen
The sum of the amounts of the nitric oxide and nitrogen dioxide contained in a gas sample calculated as if the nitric oxide were in the form of nitrogen dioxide.

oxygen (O)
Atomic number 8. The most abundant element in the Earth's crust, it also comprises about one fifth of the atmosphere.

ozone (O₃)
An allotropic form of oxygen found mainly in the ozone layer in the stratosphere. It is a bluish gas which readily absorbs ultraviolet radiation.

P

Pan
The international radiotelephony urgency signal. When repeated three times indicates uncertainty or alert, followed by nature of urgency.

parabola
A curve traced by a point equidistant from a focal point and a straight line. Its equation is $y^2 = 4ax$ where a is the distance from origin to focus.

parabolic reflector
Used for producing a parallel beam of light (searchlight) or a beam of microwave radiation (dish aerial).

parachute
A device used, or intended to be used, to retard the fall of a body or object through the air.

paraffin
A mixture of hydrocarbons. Boiling points between 150°C and 300°C. Also called kerosine.

parallax error
The reading error resulting from viewing an instrument or display from other than head on.

parallel
The intersection of the earth's surface with a plane parallel to the equator.

parallel ILS approaches
ILS approaches to parallel runways by IFR aircraft that, when established inbound towards the airport on the adjacent localiser courses, and radar-separated by at least two miles. See also SIMULTANEOUS ILS APPROACHES.

parallel offset route
A parallel track to the left or right of the designated or established airway/route. Normally associated with area navigation (RNAV) operations. See also AREA NAVIGATION.

parallelogram
A four-sided figure having its opposite sides parallel, used for calculating the effect of two forces or two velocities acting in different directions.

parallel resistances
The combined resistance resulting from several resistors connected in parallel is called the 'equivalent resistance' and is calculated by the formula

$$\frac{1}{R} = \frac{1}{r_1} + \frac{1}{r_2} + \frac{1}{r_3} \text{ etc.}$$

parallel runways
Two or more runways at the same airport whose centrelines are parallel. In addition to runway number, parallel runways are designated as L (left) and R (right) or, if three parallel runways exist, L (left), C (centre) and R (right).

parallel standard
One of the two standard parallels of a Lambert projection along which all distances are true; a parallel of true scale in any projection.

parcel
A small volume of air, small enough to contain uniform distribution of its meteorological properties, and large enough to remain relatively self-contained and respond to all meteorological processes. No specific dimensions have been defined. However, the order of magnitude of one cubic foot has been suggested.

parity (computers)
A condition where the sum of all the bits in an array of bits satisfies a nominated numerical criterion.
Note: If the numerical criterion is such that the sum must be an even number, an array of bits which satisfies it is then said to have even parity. If the criterion is such that the sum must be an odd number, an array of bits which satisfies it is then said to have odd parity.

parity error
A situation where a parity criterion is not satisfied.

partial obscuration
A designation of sky cover when part of the sky in hidden by surface-based obscuring phenomena.

pascal
Unit of pressure = 1 newton per m^2.

passenger protective breathing equipment
When fitted, the equipment in the form of a smoke hood is designed to protect the eyes, nose and mouth of the wearer from the effects of toxic and irritant smoke and fumes produced in an aircraft fire.

pavement classification number
A number expressing the bearing strength of a pavement for unrestricted operations.

P-code (satellites)
The precise or protected code. A very long sequence of pseudo-random binary biphase modulations on the global positioning system (GPS) carrier at a chip rate of 10.23 MHz which repeats about every 267 days. Each one week segment of this code is unique to one GPS satellite and is reset each week.

pendulum
A weight suspended by a wire or string. The time (T) taken for one swing = $2 \pi \sqrt{1/g}$ where 1 is the length of the wire and g is gravity. Neither the weight nor the length of the swing affects the time.

percussion cap
Usually contains mercuric fulminate, which explodes when struck, to detonate the main charge of explosive in fire-arms or bombs.

perimeter
The distance around the sides of an area.

peripherals
Units or devices that operate in conjunction with a computer but are not part of it; more generally, units or systems connected to a system under consideration but not part of the system.

permanent echo
Radar signals reflected from fixed objects on the earth's surface, e.g. buildings, towers, terrain. Permanent echoes are distinguished from 'ground clutter' by being definable locations rather than large areas. Under certain conditions they may be used to check radar alignment.

Permit to Fly [UK]
Issued by the CAA for those aircraft that do not qualify for a Certificate of Airworthiness.

perpendicular
At right angles; one line at 90° to another.

personnel locator beacon
A lightweight radiobeacon carried on the person, or in the cockpit, and transmitting a distinctive audio signal for homing, in case of emergency.

petrochemicals
Substances derived from petroleum or natural gas.

petrol
A mixture of hexane, heptane octane and other hydrocarbons. Also called gasoline.

petroleum
Natural crude oil. Also called mineral oil.

phantom beacon
A waypoint, in an RNAV system based on VOR/DME, at which no actual radio beacon exists, its position being defined in terms of bearing and distance from the nearest in-range beacon.

photo-electric effect
The transfer of energy from light (photons) falling on a substance to electrons inside the substance. Usually the emission of electrons from the substance.

photo reconnaissance
Military activity that requires locating individual photo targets and navigating to the targets at a pre-planned angle and altitude.

photosensitive
A device which changes its electrical characteristics when exposed to light, for example photocell, photodiode, phototransistor.

pictorial display and course line computer
An airborne instrument which displays the position of an aircraft as a tiny aircraft (or other symbol) moving over a chart. It also permits the tracking of parallel or offset courses and, in effect, enables a pilot to establish a 'phantom VOR' which may be used in the same way as a real VOR, at any desired position within range.

pilotage
Navigation by visual reference to landmarks.

pilot balloon
A small balloon which indicates the movements of the air aloft by its drift, as observed from the ground.

pilot balloon observation
A method of winds-aloft observation by visually tracking a pilot balloon (commonly called PIBAL).

pilot briefing [US]
A service provided by a Flight Service Station to assist pilots in flight planning. Briefing items may include weather information, NOTAMs, military activities, flow control information and other items as requested.

pilot controller system
Air–ground radiotelephony facilities implemented primarily to provide a means of direct communication between pilots and controllers.

pilot-in-command
The pilot responsible for the operation and safety of the aircraft during flight time.

pilots' automatic telephone weather answering service [US]
A continuous telephone recording containing current and forecast weather information for pilots. Known as Airmet in the United Kingdom.

pilot's discretion
When used in conjunction with altitude assignments, means that ATC has offered the pilot the option of starting climb or descent whenever he wishes and conducting the climb or descent at any rate he wishes. He may temporarily level off at any intermediate altitude. However, once he has vacated an altitude he may not return to that altitude.

pilot (to)
To manipulate the flight controls of an aircraft during flight time.

pilot weather report
A report of meteorological phenomena encountered by aircraft in flight and reported by the pilot.

pitching moment
The twisting effect applied by a system of forces acting in a pitching motion about a selected centre.

pitch setting
The propeller blade setting as determined by the blade angle measured in a manner, and at a radius, specified by the instruction manual for the propeller.

plane of the nominal ILS glide path
A plane perpendicular to the vertical plane of the runway centreline extended and containing the nominal ILS glide path.

plan position indicator (scope)
A radio indicator scope displaying range and azimuth of targets in polar co-ordinates.

plastics
Synthetic substances that are stable (solid) at normal temperatures but which have been moulded into shape by temperature and pressure.

platinum (Pt)
Atomic number 78, relative density 21.45, melting point 1773°C. Its coefficient of expansion is 9×10^{-6} per °C which is the same as glass; this makes it useful in some aspects of instrument making.

plow wind [US]
The spreading downdraft of a thunderstorm; a strong, straight-line wind in advance of the storm.

plutonium (Pu)
Atomic number 94, relative density 19.84, melting point 639.5°C. Used in nuclear weapons. One kilogram is equivalent to 10^{14} Joules (20 kilotons of TNT).

pneumatic
Filled with, or using, compressed air.

point of equal time
The point from which the time to return is equal to the time to reach the destination.

point of no return
The point beyond which an aircraft has insufficient fuel to return to the starting point.

polar air
An air mass with characteristics developed over high latitudes, especially within the sub-polar highs. Continental polar air (cP) has cold surface temperatures, low moisture content, and, especially in its source regions, has great stability in the lower layers. It is shallow in comparison with arctic air. Maritime polar (mP) initially possesses similar properties to those of continental polar air, but in passing over warmer water it becomes unstable with a higher moisture content.

polar diagram
A plot of points of equal field strength which gives a diagrammatic representation of the directional properties of an antenna.

polar front
The semi-permanent, semi-continuous front separating air masses of tropical and polar origins.

pole geographic
The intersection of the earth's axis with the surface of the earth.

pole magnetic
A rather large area on the earth where a freely suspended magnetic needle would point vertically. The north magnetic polar area is located about 15° from the geographic pole, in the vicinity of the meridian of 101° west longitude; the south magnetic polar area is about 23° from the geographic pole near longitude 143° east. The magnetic poles are in no sense 'points', but are indefinite areas, several hundred miles in extent.

Popular Flying Association
The representative body in the United Kingdom of amateur constructors and operators of ultra light and group operated aircraft.

positional response
That element of an SSR response which represents the actual position of the associated aircraft on the display.

position report
A report over a known location as transmitted by an aircraft to ATC.

position symbol
A computer generated indication shown on a radar display to indicate the mode of tracking.

positive control
The separation of all air traffic, within designated airspace, by air traffic control.

positive control airspace [US]
Space reserved wholly for IFR flight under air traffic control, it is all the airspace above 18 000 ft over approximately the north-east quarter of the United States. There are also minimum requirements as to flight instruments and pilot training. In this airspace positive separation is provided from other aircraft in flight.

potential difference
The difference between the positive and negative poles of a battery when no current is flowing. It is measured in volts, as in emf when a current is flowing.

potential energy
The energy available to a body because of its position, e.g. a coiled spring, or a glider at altitude. The potential energy of the glider is mass × gravity × altitude.

poundal
A unit of force, approximately $\frac{1}{32}$ lb. 1 pdl = 0.13825 N.

pound weight
Used also as a unit of mass = 0.45359237 kg.

power
Measured in watts (745.7 watts = 1 horsepower), it is the rate of doing work.

power density
In radar meteorology, the amount of radiated energy per unit cross sectional area in the radar beam.

power-unit
A system of one or more engines and ancillary parts which are together necessary to provide thrust, independently of the continued operation of any other power-unit(s), but not including short period thrust-producing devices.

practice instrument approach
An instrument approach procedure conducted by an aircraft, in VMC or IMC for the purpose of pilot training or proficiency demonstrations.

precipitation
Any or all forms of water particles (rain, sleet, hail or snow), that fall from the atmosphere and reach the surface.

precise positioning service
The most accurate dynamic positioning possible with the global positioning system (GPS), based on the dual frequency P-code.

precision approach
An instrument approach to landing using precision azimuth and glidepath guidance with minima as determined by the category of operation.

precision approach path indicator system
Units which direct a beam of light, red in the lower half and white in the upper, towards the approach. They are set at different elevation angles so as to give a combination of red and white for an on-slope signal, all-red if the pilot is too low, and all-white if he is too high. A transition zone will be perceived between the red and white sectors of each unit. PAPI units have a perceived transition of approximately 3 minutes of arc and are known as sharp transition units.

precision approach procedure
Based on the use of an ILS (including electronic glide slope) or Precision Approach Radar (PAR).

precision approach radar [ICAO]
A ground-based primary radar equipment used to determine the position of an aircraft during final approach, in terms of lateral and vertical deviations relative to a nominal approach path and in range relative to touchdown.

precision approach radar
Radar equipment in some ATC facilities at civil/military locations and separate military installations to detect and display azimuth, elevation and range of aircraft on the final approach course to a runway. This equipment may be used to monitor certain non-radar approaches, but is primarily used to conduct a precision instrument approach (PAR) wherein the controller issues guidance instructions to the pilot based on the aircraft's position in relation to the final approach course (azimuth), the glide path (elevation), and the distance (range) from the touchdown point on the runway as displayed on the radar scope. The abbreviation PAR is also

used to denote preferential arrival routes in ARTCC computers. See also SLIDEPATH and PREFERENTIAL ROUTES.

precision approach runway [UK]
A runway intended for the operation of aircraft using visual and non-visual aids providing guidance in both pitch and azimuth adequate for a straight-in approach. These runways are divided into three categories as follows:

(1) Category I. Intended for operations down to 200 ft decision height and down to an RVR of the order of 800 m.
(2) Category II. Intended for operations down to 100 ft decision height and down to an RVR of the order of 400 m.
(3) Category III. A. Intended for operations down to RVR of the order of 200 m.
                B. Intended for operations down to an RVR of the order of 50 m.
                C. Intended for operations without reliance on external visual reference.

preferential routes [US]
Preferential routes are adapted in ARTCC computers to accomplish inter/intra facility controller co-ordination and to assure that flight data is posted at the proper control positions. Locations having a need for these specific inbound and outbound routes normally publish such routes in local facility bulletins and their use by pilots minimises flight plan route amendments. When the work load or traffic situation permits, controllers normally provide radar vectors or assign requested routes to minimise circuitous routing. Preferential routes are usually confined to one ARTCC's area and are referred to by the following names or acronyms:

(1) Preferential departure route (PDR). A specific departure route from an airport or terminal area to an en-route point where there is no further need for flow control. It may be included in a standard instrument departure (SID) or a preferred IFR route.
(2) Preferential arrival route (PAR). A specific arrival route from an appropriate en-route point to an airport or terminal area. It may be included in a standard terminal arrival route (STAR) or a preferred IFR route. The abbreviation PAR is used primarily within the ARTCC and should not be confused with the abbreviation for precision approach radar.
(3) Preferential departure and arrival route (PDAR). A route between two terminals that are within or immediately adjacent to one ARTCC's area. They are not synonomous with preferred IFR routes but may be listed as such as they accomplish essentially the same purpose.

preferred IFR routes [US]
Routes established between busier airports to increase system efficiency and capacity. They normally extend through one or more ARTCC areas and are designed to achieve balanced traffic flows within high-density terminals. IFR clearances are issued on the basis of these routes except when severe weather avoidance procedures or other factors dictate otherwise. Preferred IFR routes are listed in the *Airport/Facility Directory*. If a flight is planned to or from an area having such routes but the departure or arrival point is not listed in the *Airport/Facility Directory*, pilots mays use that part of a preferred IFR route that is appropriate for

the departure or arrival point that is listed. Preferred IFR routes are correlated with SIDs and STARs and may be defined by airways, jet routes, direct routes between navaids, waypoints, navaid radials/DME or any combinations thereof. See also PREFERENTIAL ROUTES, STANDARD INSTRUMENT DEPARTURE and STANDARD TERMINAL ARRIVAL ROUTE.

preferred route(s)
Route(s) between given points or areas of origin and destination which, primarily for economic reasons, are identified by aircraft operators as preferable when compared with other possible or available routings.

pre-flight pilot briefing [US]   See PILOT BRIEFING.

pre-ignition
Normal combustion takes place before the sparking plug fires. This condition is often due to localised hot spots, e.g. carbon deposits becoming incandescent and igniting the mixture. Although pre-ignition can be the result of detonation, it can also be caused by high power operation in lean mixture. Pre-ignition is usually indicated by engine roughness, backfiring or a sudden increase in cylinder head temperature. The best 'in flight' measures for correcting detonation or pre-ignition are to enrich the mixture, decrease power and/or open the engine cowl flaps (if available).

presbyopia
A state of long-sightedness which afflicts an individual later in life and is usually noticed for the first time in the early forties.

pressure
Force divided by area, measured in pascals (Nm^{-2}), pound per square inch, bars, atmospheres and both inches and millimetres of Hg.

pressure altimeter
An aneroid barometer with a scale graudated in altitude instead of pressure using standard atmospheric pressure–height relationships. Shows indicated altitude (not necessarily true altitude). May be set to measure altitude (indicated) from any arbitarily chosen level. See also ALTIMETER SETTINGS, ALTITUDE.

pressure altitude
An atmospheric pressure expressed in terms of altitude which corresponds to that pressure in the standard atmosphere. Since an altimeter operates solely on pressure, this is the uncorrected altitude indicated by an altimeter set at standard sea level pressure of 29.92 inches or 1013 millibars (hectopascals).

pressure gradient
The change in atmospheric pressure per unit of horizontal distance.

pressure jump (Meteorology)
A sudden, significant increase in station pressure.

prevailing easterlies
The broad current or pattern of persistant easterly winds in the tropics and in polar regions.

prevailing visibility [US]
The greatest horizontal visibility which is equalled or exceeded throughout half of the horizon circle (it need not be a continuous half).

prevailing westerlies
The dominant west-to-east motion of the atmosphere, centred over middle latitudes of both hemispheres.

prevailing wind
Direction from which the wind blows most frequently.

preventive maintenance
Simple or minor preservation operations and the replacement of small standard parts not involving complex assembly operations.

primary area
A defined area symmetrically disposed about the nominal flight track in which full obstacle clearance is provided.

primary coil
The input coil in a transformer or an induction coil.

primary frequency
The radiotelephony frequency assigned to an aircraft as a first choice for air–ground communication in a radiotelephony network.

primary means of communications
The means of communication to be adopted normally by aircraft and ground stations as a first choice where alternative means of communication exist.

primary radar
A radar system which uses reflected radio signals.

primary runway(s)
Runway(s) used in preference to others whenever conditions permit.

principal point
The centre of a vertical photograph found by joining the collimating marks.

printed circuit
An electronic circuit in which all the wiring and most of the components are printed onto an insulated base; the 'silicon chip'.

printed communications
Communications which automatically provide a permanent printed record at each terminal of a circuit of all messages which pass over such circuit.

probability
In relation to meteorological forecasts means the probability of occurrence as a percentage (never more than 50%).

procedure turn
A turn made (usually) at the standard rate of 3° a second (360° in 2 minutes), in connection with an instrument approach. The manoeuvre prescribed when it is

necessary to reverse direction to establish an aircraft on the intermediate approach segment or final approach track. The outbound track, direction of turn, distance within which the turn must be completed and minimum altitude are specified in the procedure. However, unless otherwise restricted, the point at which the turn may be commenced and the type and rate of turn are left to the discretion of the pilot.

procedure turn [ICAO]
A manoeuvre in which a turn is made away from a designated track followed by a turn in the opposite direction, both turns being executed so as to permit the aircraft to intercept and proceed along the reciprocal of the designated track. Procedure turns are designated left or right according to the direction of the initial turn. They may also be designated as being made either in level flight or while descending, according to the circumstances of each individual instrument approach procedure.

prodat
An experimental system developed by the European Space Agency to provide low rate data communications with mobiles via satellite.

profile
The orthogonal projection of a flight path or portion thereof on the vertical surface containing the nominal track.

profile descent
An uninterruped descent (except where level flight is required for speed adjustment, e.g. 250 knots at 10 000 ft amsl) from cruising altitude/level to interception of a glideslope or to a minimum altitude specified for the initial or intermediate approach segment of a non-precision instrument approach. The profile descent normally terminates at the approach gate or where the glideslope or other appropriate minimum altitude is intercepted.

prognostic chart
A forecast of a specified meteorological element(s) for a specified time or period and a specified surface or portion of airspace, depicted graphically on a chart.

programmable indicator data processor [US]
A modification to the AN/TPX–42 interrogator system currently installed in fixed rapcons. It detects, tracks and predicts secondary radar aircraft targets. These are displayed by means of computer-generated symbols and alphanumeric characters depicting flight identification of aircraft altitude, groundspeed and flight plan data. Although primary radar targets are not tracked, they are displayed coincident with the secondary radar targets as well as with the other symbols and alphanumerics. The system has the capability of interfacing with ARTCCs.

program
A logical, step-by-step series of instructions fed into a computer.

progress report   See POSITION REPORT.

prohibited area
An airspace of defined dimensions, above the land areas or territorial waters of a State, within which the flight of aircraft is prohibited.

projectile
Something which is thrown or shot (not a rocket-propelled missile). If air resistance is ignored, then, with velocity (v) and angle from horizontal (a):

maximum height $= (v^2 \sin^2 a)2g$
time to reach maximum height $= (v \sin a)/g$
maximum range $= (v^2 \sin 2a)/g$
total time of flight $= (2v \sin a)/g$

projection
The system of reference lines representing the earth's meridians and parallels on a chart. The projection is usually designed to retain some special property of the sphere, such as true directions, true distances, true shape or true area.

propane (C_3H_8)
Inflammable gas used as a fuel. Boiling point 42.17°C.

propellant
Explosive chemicals used to fill cartridges, and shell cases; rocket fuels and their oxidants; gas used in aerosol cans.

propeller
A device for propelling an aircraft that has blades on an engine-driven shaft and that, when rotated, produces by its action on the air a thrust approximately perpendicular to its plane of rotation. It includes control components normally supplied by its manufacturer, but does not include main and auxiliary rotors or rotating aerofoils of engines.

propeller efficiency
The effectiveness of a propeller, expressed in terms of the ratio thrust horsepower to shaft horsepower.

proportional guidance sector
The volume of airspace within which the angular guidance information provided by a function is directly proportional to the angular displacement of the airborne antenna with respect to the zero angle reference.

proposed boundary crossing time
Each ATC centre has this PBCT parameter for each internal airport. Proposed internal flight plans are transmitted to the adjacent centre if the flight time along the proposed route from the departure airport to the centre boundary is less than or equal to the value of PBCT, or if airport adaptation specifies transmission regardless of PBCT.

propulsion system
A system consisting of a power-unit and all other equipment utilised to provide those functions necessary to sustain, monitor and control the power/thrust output of any one power-unit following installation on the airframe.

proton
A positively charged particle found in the nucleus with a mass 1836.12 times greater than the electron.

protractor
A device for measuring angles on charts and diagrams. Usually associated with a straight edge for drawing the angle measured.

pseudolite
A ground-based differential GPS receiver which transmits a signal like that of an actual GPS satellite and can be used for ranging.

pseudo-random code
A signal with random noise like properties. It is complicated but repeats a pattern of 1s and 0s.

pseudorange
A distance measurement based on the correlation of a satellite-transmitted code and the local receiver's reference code, that has not been corrected for errors in synchronisation between the transmitter's clock and the receiver's clock.

psychrometer
An instrument for measuring atmospheric humidity, consisting of a dry-bulb thermometer and wet-bulb thermometer (covered with a muslin wick). Used in the calculation of dew point and relative humidity.

public aircraft [US]
Aircraft used only in the service of a government or a political sub-division. It does not include any government-owned aircraft engaged in carrying persons or property for commercial purposes.

public authorities
The agencies of officials of an ICAO Contracting State responsible for the application and enforcement of the particular laws and regulations of that State which relate to any aspect of ICAO standards and recommended practices.

public transport [UK]
Public transport shall have the meaning as defined in the Air Navigation Order. This includes the carriage for hire and reward of passengers and cargo.

published route
A route for which an IFR altitude has been established and published, e.g. airways, advisory routes, area navigation routes, specified direct routes.

pulse
A brief increase in voltage or current. Pertaining to radar, a brief burst of electromagnetic radiation emitted by the radar, of very short time duration. See also PULSE LENGTH.

pulse amplitude
(1) Loran-A. The peak voltage of the pulse envelope.
(2) SSR. The peak voltage amplitude of the pulse.
(3) DME. The maximum voltage of the pulse envelope.

pulse code
The method of differentiating between W, X, Y and Z modes.

pulse decay time
(1) DME. The time as measured between the 90% and 10% amplitude points on the trailing edge of the pulse envelope.
(2) SSR. The decay time as measured between 0.9A and 0.1A on the trailing edge of the pulse envelope.

pulse duration
(1) DME. The time interval between 50% amplitude point on leading and trailing edges of the pulse envelope.
(2) SSR. The time interval between 0.5A points on leading and trailing edges of the pulse envelope.

pulse interval (SSR)
The time interval between the 0.5A point on the leading edge of the first pulse and the 0.5A point on the leading edge of the second pulse.

pulse jet
In this engine, the forward movement of the aircraft forces air through the intake valve into the combustion chamber. Injected fuel is then fired, but the intake valve does not open again until the pressure in the combustion chamber is lower than that of the forward movement. The thrust comes from this rapid series of separate explosions. See also RAM JET.

pulse length
Pertaining to radar, the dimension of a radar pulse. May be expressed as the time duration or the length in linear units. Linear dimension is equal to time duration multiplied by the speed of propagation (approximately the speed of light).

pulse repetition period (Loran-A)
The time interval (microseconds) between the 0.5A point on the leading edge of the first pulse and the 0.5A point on the leading edge of the following pulse from the same station.

pulse rise time
(1) DME. The time as measured between the 10% and 90% amplitude points of the leading edge of a pulse envelope,
(2) Loran-A. The time interval (microseconds) between 0.1A point and 0.9A point on the leading edge of a pulse envelope.
(3) SSR. The rise time as measured between 0.1A and 0.9A on the leading edge of a pulse envelope.

pulse width (Loran-A)
The time interval (microseconds) between 0.5A points on leading and trailing edges of a pulse envelope.

pyrene
A trade name for carbon tetrachloride ($C Cl_4$) used for fire extinguishers.

pyrometers
Thermometers for measuring high temperatures:

(1) platinum resistance thermometers.
(2) thermocouples.
(3) optical pyrometers.
(4) radio-micrometers.

pyrophoric liquid
A liquid which may ignite spontaneously when exposed to air, the temperature of which is 55°C or below.

pyrotechnics
Pertaining to fireworks. Used for warning unauthorised aircraft that they are overflying prohibited areas. Also used by aerodrome ATC to inform pilots that conflict exists either on the approach or on the ground.

pyrotechnic substance
A mixture or compound designed to produce an effect by heat, light, sound, gas or smoke, or a combination of these, as a result of non-detonative self-sustaining exothermic chemical reactions.

Pythagoras's theorem
In a right-angled triangle, the square of the length of the hypotenuse is equal to the sum of the squares of the lengths of the other two sides.

Q-code
An international telegraph code, originally intended to accelerate commercial W/T operation, which has since proved so convenient that it is also used in RTF operation. Typical examples:

(1) QFE. The atmospheric pressure at airfield level is . . . .
(2) QTE. Your true bearing from this station is . . . .

quadrant
A quarter part of a circle centred on a navaid, oriented clockwise from magnetic north as follows:

(1) NE quadrant 000 – 089.
(2) SE quadrant 090 – 179.
(3) SW quadrant 180 – 269.
(4) NW quadrant 270 – 359.

quadrantal cruising levels
Specified cruising levels determined in relation to magnetic track within quadrants of the compass.

quadrantal error
A signal error caused by the presence of a metal structure, e.g. the receiving aircraft.

quadrantal heights   See QUADRANTAL CRUISING LEVELS.

quadratic equation
$ax^2 + bx + c = 0$
This may be solved by the formula:

$$x = \frac{-b \pm \sqrt{(b^2 - 4ac)}}{2a}$$

The sum of the roots $= -b/a$

quadrilateral
A four-sided figure.

quart
A quarter of a gallon.

quarter chord
Locus of all points lying at 25% of the chord of a wing (or other aerofoil), each measurement being in a plane parallel to the longitudinal axis of the aircraft.

quartz clock

A clock regulated by the resonance frequency of the vibration of a quartz crystal in an electric field.

quasi-stationary front

A front which is stationary or nearly so, conventionally, a front which is moving at a speed of less than 5 knots is generally considered to be quasi-stationary (commonly called a stationary front).

quota flow control

A flow control procedure by which the central flow control facility (CFCF) restricts traffic to the ARTC centre area having a congested airport, thereby avoiding sector/area saturation.

R

racetrack procedure
A procedure designed to enable the aircraft to maintain a holding pattern, level or while reducing altitude, during the initial approach segment and/or establish the aircraft inbound when the entry into a reversal procedure is not practical.

racon
Term for 'radar beacon'. It is a fixed transponder, transmitting only when triggered by a suitably equipped interrogator.

rad
A unit of energy imparted by conveying radiation to a mass of matter corresponding to 10^{-2} Joules per kg.

radar
Radio detection and ranging. Centimetre wavelengths at a radio frequency are pulse modulated and transmitted. The beams are reflected by solid objects and these echoes travel back to the transmitter/receiver and are transformed by a cathode ray tube into a visual display showing direction and distance. It is the basic principal of the radio altimeter, of DME and of a transponder. It is also employed in collision prevention systems and may be used to obtain a radarscope 'picture' of the terrain; and for other related functions. In primary radar, a small part of the radio pulse is reflected back to the transmitting station, as in ASR, ARSR and PAR; in secondary surveillance radar (SSR), the transmitter triggers a second transmission from a transponder. Since hydrometeors can scatter radio energy, weather radars, operating on certain frequency bands, can detect the presence of precipitation, clouds, or both.

radar advisory
The provision of advice and information based on radar observations. See also
ADVISORY SERVICE.

Radar Advisory Service [UK]
An air traffic radar service in which the controller will provide advice necessary to maintain standard separation between participating aircraft, and in which he will pass to the pilot the bearing, distance and, if known, level of conflicting non-participating traffic, together with advice on action necessary to resolve the confliction. Where time does not permit this procedure to be adopted, the controller will pass avoiding action to resolve the confliction followed by information on the conflicting traffic.

radar altitude
The altitude of an aircraft determined by radar-type radio altimeter, i.e. the actual distance from the nearest terrain or water feature encompassed by the downward

directed radar beam. For all practical purposes, it is the 'actual' distance above a ground or inland water surface or the true altitude above an ocean surface.

radar approach
An instrument approach procedure that utilises precision approach radar or airport surveillance radar.

radar approach [ICAO]
An approach, executed by an aircraft, under the direction of a radar controller.

radar arrival
An arriving aircraft that is being vectored to the final approach course for an instrument approach, or for a visual approach to the airport.

radar beam
The focused energy radiated by radar similar to a flashlight or searchlight beam.

radar blip
A generic term for the visual indication, in non-symbolic form, on a radar display of the position of an aircraft obtained by primary or secondary radar.

radar clear range
A firing range surveyed by radar within which the operating authority accepts responsibility for with-holding fire if an aircraft is within the area into which, and through which, missiles are liable to fall.

radar clutter
The visual indication on a radar display of unwanted signals.

radar contact
(1) Used by ATC to inform an aircraft that it is identified on the radar display and radar flight following will be provided until radar identification is terminated. Radar service may also be provided within the limits of necessity and capability.
(2) A term used to inform the controller that the aircraft is identified and approval is granted for the aircraft to enter the receiving controller's airspace.

radar contact [ICAO]
The situation that exists when the radar blip of a particular aircraft is seen and identified on a radar display.

radar contact lost
Used by ATC to inform a pilot that radar identification of his aircraft has been lost. The loss may be attributed to several things including the aircraft merging with weather or ground clutter, the aircraft flying below radar line of sight, the aircraft entering an area of poor radar return, or a failure of the aircraft transponder or ground radar equipment.

radar control
Term used to indicate that radar-derived information is employed directly in the provision of air traffic control service.

radar controller
A qualified air traffic controller holding a radar rating appropriate to the functions to which he is assigned.

radar departure
The control of a departing aircraft by the use of surveillance radar to assist it to leave the vicinity of an aerodrome safely and expeditiously.

radar display
An electronic display of radar-derived information depicting the position and movement of aircraft.

radar echo
The visual indication on a radar display of a radar signal reflected from an object.

radar flight following
The observation of the progress of radar identified aircraft, whose primary navigation is being provided by the pilot, wherein the controller retains and correlates the aircraft identity with the appropriate target or target symbol displayed on the radar scope. See also RADAR CONTACT and RADAR SERVICE.

radar handover
Transfer of responsibility for the control of an aircraft between two controllers using radar, following identification of the aircraft by both controllers.

radar heading
A magnetic heading given by a controller to a pilot on the basis of radar-derived information for the purpose of providing navigational guidance.

radar identification
The process of ascertaining that an observed radar target is the radar return from a particular aircraft. See also RADAR CONTACT, RADAR SERVICE.

radar identification (ICAO)
The process of correlating a particular radar blip with a specific aircraft.

radar information service [UK]
RIS is an air traffic radar service in which the controller will only provide traffic information. He will inform the pilot of the bearing, distance and, if known, the level of the conflicting traffic. No avoiding action will be offered. The pilot is wholly responsible for maintaining separation from other aircraft whether or not the controller has passed traffic information.

radar map
Information superimposed on a radar display to provide ready indication of selected features.

radar mile
The time taken for an e.m. wave to travel 1 NM and back, approximately 12.36 μs.

radar monitoring
The use of radar for the purpose of providing aircraft with information and advice relative to significant deviations from nominal flight path.

radar point out
Used between controllers to indicate radar hand-off action, where the initiating controller plans to retain communications with an aircraft penetrating the other controller's airspace and additional co-ordination is required.

radar position symbol
A generic term for the visual indication in a symbolic form, on a radar display, of the position of an aircraft obtained after digital computer processing of positional data derived from primary radar and/or SSR.

radar response or SSR response
The visual indication in non-symbolic form, on a radar display, of a radar signal transmitted from an object in reply to an interrogation.

radar route
A flight path or route over which an aircraft is vectored. Navigational guidance and altitude assignments are provided by ATC.

radar separation
The separation used when aircraft position information is derived from radar sources.

radar service
A term that encompasses one or more of the following services based on the use of radar that can be provided by a controller to a pilot of a radar-identified aircraft:

(1) Radar separation. Radar spacing of aircraft in accordance with established minima.
(2) Radar navigational guidance. Vectoring aircraft to provide course guidance.
(3) Radar monitoring. The radar flight following of aircraft, whose primary navigation is being performed by the pilot, to observe and note deviations from its authorised flight path, airway or route. When being applied specifically to radar monitoring of instrument approaches, i.e. with precision approach radar (PAR) or radar monitoring of simultaneous ILS approaches, it includes advice and instructions whenever an aircraft nears or exceeds the prescribed PAR safety limit or simultaneous ILS no transgression zone.
(4) Radar service (ICAO). Term used to indicate a service provided directly by means of radar.
(5) Radar separation (ICAO). The separation used when aircraft position information is derived from radar sources.
(6) Radar monitoring (ICAO). The use of radar for the purpose of providing aircraft with information and advice relative to significant deviations from nominal flight path.

radar service terminated
Used by ATC to inform a pilot that he will no longer be provided with any of the services that could be received while under radar contact.

radar-sonde observation
A rawinsonde observation in which winds are determined by radar tracking a balloon-borne target.

radar surveillance
Observance of the movements of aircraft on a radar display and the passing of advice and information to identified aircraft and, where appropriate, to other ATS units.

radar tracking
The act, either by a human or a computer, of following the movements of specific aircraft by means of radar, for the purpose of ensuring a continuous indication of the identity, position, track and/or height of the aircraft.
Note: In some cases, information other than radar-derived information is used to assist the tracking processes.

radar track position
An extrapolation of aircraft position by the computer based upon radar information and used by the computer for tracking purposes.

radar unit
That element of an air traffic services unit which uses radar equipment to provide one or more services.

radar vectoring
Provision of navigational guidance to aircraft in the form of specific headings, based on the use of radar.

radar weather echo intensity levels [US]
Existing radar systems cannot detect turbulence. However, there is a direct correlation between the degree of turbulence and other weather features associated with thunderstorms and the radar weather echo intensity. The National Weather Service has categorised six levels of radar weather echo intensity. The following list gives the weather features likely to be associated with these levels during thunderstorm weather situations:

(1) Level 1 (weak) and level 2 (moderate). Light to moderate turbulence is possible with lightning.
(2) Level 3 (strong). Severe turbulence possible, lightning.
(3) Level 4 (very strong). Severe turbulence likely, lightning.
(4) Level 5 (intense). Severe turbulence, lightning, organised wind gusts. Hail likely.
(5) Level 6 (extreme). Severe turbulence, large hail, lightning, extensive wind gusts and turbulence.

radial
One of a set of straight half lines terminating at a fixed point; a line of radio bearing from a VOR station.

radian
The angle subtended at the centre of a circle by an arc of equal length to the radius: $57.296°$. $360° = 2\pi$ rad.

radiant energy
In space, in the absence of matter, electromagnetic radiation is the only form of energy.

radiation
Travel of electromagnetic waves (at 186 000 miles per second), many of which may be visible as light. Cosmic rays, gamma rays, X-rays, ultra-violet rays, visible light rays, infra-red rays and radio waves are some common types of radiation.

radiation fog
Fog characteristically resulting when radiational cooling of the earth's surface lowers the air temperature near the ground to, or below, its initial dew point, on calm, clear nights.

radiation sickness
Caused by exposure to harmful radiation. Vomiting and diarrhoea may be followed by leukaemia in severe cases.

radio altimeter
A device for measuring the height of an aircraft above the surface of the earth (not above mean sea level), by means of reflected radio waves.

radio beacon
A non-directional radio facility such as an H beacon (homing beacon) or compass locator.

radio bearing
The angle between the apparent direction of a definite source of emission of electromagnetic waves and a reference direction, as determined at a radio direction-finding station. A true radio bearing is one for which the reference direction is that of true north. A magnetic radio bearing is one for which the reference direction is that of magnetic north.

radio direction finder
A device employing a loop antenna for the determination of the direction of radio bearings. Signals may be received aurally or visually.

radio direction finder (automatic)
Similar to the ordinary radio direction finder, except that automatic rotation of the loop is provided, and the indicator needle continuously shows the bearing of the station.

radio direction finder station
A ground station (or group of stations) equipped to determine the directional bearing (or position) of craft from the station, and reporting the results to the navigator by radio.

radio direction finding
Radiodetermination using the reception of radio waves for the purpose of determining the direction of a station or object.

radiolocation
The detection of ships, aircraft, etc. by radar.

radio magnetic indicator
An aircraft navigational instrument coupled with a gyro compass or similar compass that indicates the direction of a selected navaid and indicates bearing with respect to the heading of the aircraft.

radiosonde
A device carried aloft by a balloon equipped with measuring instruments that automatically converts temperature, pressure and humidity data into electrical impulses and transmits this information to a ground recorder.

radiotelephony network
A group of radiotelephony aeronautical stations which operate on, and guard frequencies from, the same family and which support each other in a define manner to ensure maximum dependability of air–ground communications and dissemination of air–ground traffic.

radius of action
The distance or time an aircraft may safely fly toward its destination before turning back to the starting point or to some alternate airport.

radome
A detachable aircraft nose cone made of dielectric material; more generally, a dielectric panel or antenna cover.

rain
A form of precipitation. Drops are larger than drizzle and fall in relatively straight, although not necessarily vertical, paths as compared to drizzle, which falls in irregular paths.

rainbow
The colours which are produced by the refraction of sunlight by raindrops.

rainfall
A term sometimes synonymous with rain, but most frequently used in reference to amounts of precipitation (including snow, hail, etc.).

ram jet
Called a 'flying drainpipe' and used to propel 'doodlebugs' (V1 rocket bombs) during World War II. Motor-activated by high speed launching. See also PULSE JET.

random area navigation routes
Direct routes, based on area navigation capability, between waypoints defined in terms of degree/distance fixes or offset from published or established routes/ airways at specified distance and direction.

random RNAV routes   See RANDOM AREA NAVIGATION ROUTES.

range
An alignment of landmarks along a route such that the desired track may be made good by flying so as to keep the objects continually in line.

range–height indicator scope
A radar indicator scope displaying a vertical cross-section of targets along a selected azimuth.

rapid intervention vehicle
The main function of an RIV is to reach accident sites quickly, carrying personnel able to initiate rescue action, and to provide an effective means of fire suppression pending the arrival of the major units. It should be capable of crossing adverse terrain where access for the major units may be slow or difficult. The design combines speed, acceleration, flotation, traction and manoeuvrability as far as possible, bearing in mind that these characteristics are not necessarily compatible. Speed and acceleration are considered to have preference at most aerodromes. The requirement is for the vehicle to carry an effective quantity of primary extinguishing media appropriate to the airfield category. The vehicle also carries a quantity of complementary media.

rarefaction
The opposite of compression; reduction of pressure.

rascal
Programs for the theoretical analysis of mono and multi radar cover in the ATC environment.

raster
The pattern of lines that scan the screen of a cathode-ray tube.

rated air traffic controller
An air traffic controller holding a licence and valid ratings appropriate to the privileges exercised by him.

rated coverage
The area surrounding an NDB within which the strength of the vertical field of the ground wave exceeds the minimum value specified for the geographical area in which the radio beacon is situated.
Note: The foregoing definition is intended to establish a method of rating radio beacons by the normal coverage to be expected in the absence of sky wave transmission and/or anomalous propagation from the radio beacon concerned or interference from other LF/MF facilities, but taking into account the atmospheric noise in the geographical area concerned.

rated maximum continuous augmented thrust [US]
With respect to turbojet engine type certification, the approved jet thrust that is developed statically or in flight, in standard atmosphere at a specified altitude, with fluid injection or with the burning of fuel in a separate combustion chamber, within the engine operating limitations and approved for unrestricted periods of use.

rated maximum continuous power [US]
With respect to reciprocating, turbopropeller, and turboshaft engines, the approved brake horsepower that is developed statically or in flight, in standard atmosphere at a specified altitude, within the engine operating limitations and approved for unrestricted periods of use.

rated maximum continuous thrust [US]
With respect to turbojet engine type certification, the approved jet thrust that is developed statically or in flight, in standard atmosphere at a specified altitude, without fluid injection and without the burning of fuel in a separate combustion chamber, within the engine operating limitations.

rated output
For engine emissions purposes, the maximum power/thrust available for take-off under normal operating conditions at ISA sea level static conditions without the use of water injection as approved by the certificating authority. Thrust is expressed in kilonewtons.

rated takeoff augmented thrust [US]
With respect to turbojet engine type certification, the approved jet thrust that is developed statically under standard sea level conditions, with fluid injection or with the burning of fuel in a separate combustion chamber, within the engine operating limitations and limited in use to periods of not more than five minutes for takeoff operation.

rated takeoff power [US]
With respect to reciprocating, turbopropeller, and turboshaft engine type certification, the approved brake horsepower that is developed statically under standard sea level conditions, within the engine operating limitations and limited in use to periods of not more than five minutes for takeoff operation.

rated takeoff thrust [US]
With respect to turbojet engine type certification, the approved jet thrust that is developed statically under standard sea level conditions without fluid injection and without the burning of fuel in a separate combustion chamber, within the engine operating limitations and limited in use to periods of not more than five minutes for takeoff operation.

rated 30-minute power [US]
With respect to helicopter turbine engines, the maximum brake horsepower developed under static conditions at specified altitudes and atmospheric temperatures, under the maximum conditions of rotorshaft rotational speed and gas temperature, and limited in use to periods of not more than 30 minutes as shown on the engine data sheet.

rated two-and-one-half-minute power [US]
With respect to helicopter turbine engines, the brake horsepower developed statically in standard atmosphere at sea level, or at a specified altitude, for one-engine-out operation of multi-engine helicopters for two-and-one-half-minutes at rotorshaft rotation speed and gas temperature established for this rating.

rating
(1) An authorisation entered on or associated with a licence and forming part thereof, stating special conditions, privileges or limitations pertaining to such licence.
(2) A statement that, as a part of certificate, sets forth special conditions, privileges or limitations.

rawinsonde observation
A combined winds aloft and radiosonde observation. Winds are determined by tracking the radiosonde by radio direction finder or radar.

reactance (electricity)
That part of the impedance which is due to inductance or capacitance, or both, and which stores energy rather than dissipates it.

reaction propulsion
Jet propulsion based on Newton's third law of motion: 'for every action there is an equal and opposite reaction'. The high velocity high pressure stream of gas pushed backwards by the jet engine reacts by pushing the jet engine forwards. The lower the outside air pressure the more efficient the jet. The same principle applies to rockets in space.

read back (RTF)
Repeat my message back to me.

receiving controller/facility
A controller/facility receiving control of an aircraft from another controller/facility.

reciprocal
Of a number, is 1 divided by the number of a bearing, is the opposite on the compass card, i.e. the bearing plus or minus 180°C.

recommended practice [ICAO]
Any specification for physical characteristics, configuration, matériel, performance, personnel or procedure, the uniform application of which is recognised as desirable in the interests of safety, regularity or efficiency of international air navigation, and to which ICAO Contracting States will endeavour to conform in accordance with the Convention on International Civil Aviation.

rectification
Changing an alternating current into a direct current.

rectilinear
In a straight line.

re-entry
Re-entering the earth's atmosphere from space. The angle of re-entry is crucial: too steep an angle would produce too much heat; and too shallow an angle would make the spacecraft bounce off into space again.

reference humidity
The relationship between temperature and reference humidity is defined as follows:

(1) at temperatures at and below ISA, 80% relative humidity,
(2) at temperatures at and above ISA + 28°C, 34% relative humidity,
(3) at temperatures between ISA and ISA + 28°C, the relative humidity varies linearly between the humidity specified for those temperatures.

reference pressure ratio
The ratio of the mean total pressure at the last compressor discharge plane of the compressor to the mean total pressure at the compressor entry plane when the engine is developing a takeoff thrust rating in ISA sea level static conditions.

reflex angle
An angle greater than 180° and less than 360°.

refraction
In radar, bending of the radar beam by variations in atmospheric density, water vapour content and temperature.

(1) Normal refraction. Refraction of the radar beams under normal atmospheric conditions; normal radius of curvature of the beam is about four times the radius of curvature of the earth.
(2) Super-refraction. More than normal bending of the radar beam resulting from abnormal vertical gradients of temperature and/or water vapour.
(3) Sub refraction. Less than normal bending of the radar beam resulting from abnormal vertical gradients of temperature and/or water vapour.

Regional Administrative Radio Conference
A conference for the planning of the medium frequency maritime mobile and aeronautical radionavigation services.

regional air navigation agreement
Agreement approved by the Council of ICAO normally on the advice of a Regional Air Navigation Meeting.

regional area forecast centre
A meteorological centre designed to prepare and supply area forecasts for flights departing from aerodromes within its service area and to supply grid point data in digital form for up to world-wide coverage.

regular aerodrome
An aerodrome which may be listed in the flight plan as an aerodrome of intended landing.

regular station
A station selected from those forming an en-route air–ground radiotelephony network to communicate with or to intercept communications from aircraft in normal conditions.

relative bearing
The bearing of a radio station or object relative to the aircraft's heading.

relative density
The density of a substance compared to the density of water which is 1.00. It is equal to the weight in grams per cubic centimetre. Any substance with an RD of less than one will float on water. Any substance with an RD of more than one will sink.

relative humidity
The amount of water vapour in the air compared to the maximum the air could absorb at that temperature. Hot air can absorb more than cold air so, if the temperature goes up, the RH goes down; and if the temperature cools, the RH goes up until it is 100%. After that (the dew point), dew, mist, fog or cloud begins to form.

relative movement
The movement of an aircraft relative to another craft also in motion.

relay
An electrical device by which an electric current in one circuit is used to switch another circuit on or off.

release time
A departure time restriction issued to a pilot by ATC when necessary to separate a departing aircraft from other traffic.

release time (ICAO)
Time prior to which an aircraft should be given further clearance or prior to which it should not proceed in case of radio failure.

relief
The inequalities in elevation of the surface of the earth; usually represented on the aeronautical charts by contours, gradient tints or hill shading. Elevations are generally expressed in feet (or metres) above mean sea level, the one exception on aeronautical charts being that the shore line is the line of high water.

remote communications
An unmanned VHF/UHF transmitter/receiver facility which is used to expand ARTCC air–ground communications coverage and to facilitate direct contact between pilots and controllers. RCAG facilities are not always equipped with the emergency frequencies 121.5 MHz and 243.0 MHz.

remote communications outlet
An unmanned air–ground communications station remotely controlled, providing UHF and VHF transmit and receive capability to extend the service range.

remote scope
In radar meteorology, a secondary or 'slave' scope remoted from weather radar.

rendering (a certificate of airworthiness valid)
The action taken by a Contracting State, as an alternative to issuing its own Certificate of Airworthiness, in accepting a Certificate of Airworthiness issued by any other Contracting State as the equivalent of its own Certificate of Airworthiness.

rendering (a licence) valid
The action taken by a Contracting State, as an alternative to issuing its own licence, in accepting a licence issued by any other Contracting State as the equivalent of its own licence, i.e. validation.

repetitive flight plan
a flight plan related to a series of frequently recurring, regularly operated, individual flights with identical basic features, submitted by an operator for retention and repetitive use by ATS units.

replacement
A work operation which involves the removal and replacement of the same part or the substitution of a similar part.

reply efficiency
The ratio of replies transmitted by a transponder to the total of received valid interrogations.

report
Used to instruct a pilot to advise ATC of specified information, e.g. 'Report passing Daventry VOR'.

reporting point
A geographical location in relation to which the position of an aircraft is reported.

reporting point [ICAO]
A specified geographical location in relation to which the position of an aircraft can be reported.

request full route clearance [US]
Used by pilots to request that the entire route of flight be read verbatim in an ATC clearance. Such a request should be made to preclude receiving an ATC clearance based on the originally filed flight plan when a filed IFR flight plan has been revised by the pilot, company or operations prior to departure.

rescue co-odination centre
A search and rescue (SAR) facility equipped and manned to co-ordinate and control SAR operations in an area designated by the SAR plan.

rescue co-ordination centre [ICAO]
A unit responsible for promoting efficient organisation of search and rescue service and for co-ordinating the conduct of search and rescue.

rescue subcentre
A unit subordinate to a rescue co-ordination centre, established to complement the latter within a specified portion of a search and rescue region.

rescue unit
A unit composed of trained personnel and provided with equipment suitable for the expeditious conduct of search and rescue.

resistance (R)
Electrical resistance is measured in Ohms (Ω). The basic formula is ohms = volts divided by amperes but, as a result of resistance, heat is produced and is calculated by joules = amps^2 × ohms × seconds.

resolution
Pertaining to radar, the ability of radar to show discrete targets separately, i.e. the better the resolution, the closer two targets can be to each other, and still be detected as separate targets.

(1) Beam resolution. The ability of radar to distinguish between targets at approximately the same range but at different azimuths.

(2) Range resolution. The ability of radar to distinguish between targets on the same azimuth but at different ranges.

rest period

As applicable to flight crew, duty time limitation is a period of time before starting a flying duty period which is designed to give flight crew members adequate opportunity to rest before a flight. Flight time and flight duty period limitations are established for the sole purpose of reducing the probability that fatigue of flight crew members may adversely affect the safety of the flight. In order to guard against this, two types of fatigue must be taken into account, namely, transient fatigue and cumulative fatigue. Transient fatigue may be described as fatigue which is normally experienced by a healthy individual following a period of work, exertion or excitement, and it is normally dispelled by a single sufficient period of sleep. On the other hand cumulative fatigue may occur after delayed or incomplete recovery from transient fatigue or as the after-effect of more than a normal amount of work, exertion or excitement without sufficient opportunity for recuperation. Flight and duty time limitation schemes are usually a statutory requirement in ICAO Contracting States but the time periods involved will vary between countries and the different types of operations.

restricted area

An airspace of defined dimensions, above the land areas or territorial waters of a State, within which the flight of aircraft is restricted in accordance with certain specified conditions.

restricted zone [UK]

In relation to an aerodrome or air navigation installation, any part of the aerodrome or installation designated under the Aviation and Maritime Security Act or, where the whole of the aerodrome or installation is so designated, that aerodrome or installation.

resultant

A single force or velocity that produces the same effect as two or more forces acting together.

resume own navigation

Used by ATC to advise a pilot to resume his own navigational responsibility. It is issued after completion of a radar vector or when radar contact is lost while the aircraft is being radar vectored.

retardation

Decrease of velocity, deceleration, negative acceleration.

retro-rocket

A small rocket designed to fire at the appropriate time, in the opposite direction to the main propulsion system, in order to decelerate the vehicle.

reversal procedure

A procedure designed to enable aircraft to reverse direction during the initial approach segment of an instrument approach procedure. The sequence may include procedure turns or base turns.

reversal zone
A zone within an indicated course sector or an indicated ILS glide path sector in which the slope of the sector characteristic curve is negative.

Reynolds number (Re)

A non-dimentional coefficient $= \text{pVL}/\mu = \left(\dfrac{\text{Density} \times \text{Velocity} \times \text{size}}{\text{Viscosity}} \right)$ which is

used to solve problems when testing scale models in wind tunnels.

rheostat
A variable electrical resistor used for dimming and brightening lights, changing the speed of fans, etc.

rhodopsin
A substance often called visual purple and upon which the quality of a person's night vision depends.

rho-theta
Delineation of position by specifying range and bearing from a fixed point.

rib
A member which maintains the shape of a plane or aerofoil.

ridge
In meteorology, an elongated area of relatively high atmospheric pressure. Usually associated with, and most clearly identified as, an area of maximum anticyclonic curvature of the wind flow (isobars, contours or streamlines). Also called ridge line.

rigging
The assembly or relative alignment or adjustment of the various components of the aircraft.

rime
A milky and opaque granular deposit of ice formed by the rapid freezing of super-cooled water drops as they strike an exposed object. It is denser and harder than hoar frost, but lighter, softer and less transparent than glaze.

RNAV approach
An instrument approach procedure that relies on aircraft area navigation equipment for navigational guidance.

rocket
Driven by reaction propulsion. It contains its own propellants and so does not need the oxygen of the air and can be used in space. 'Rocket motors' are used on certain types of aircraft to assist takeoff.

rocketsonde
A type of radiosonde launched by a rocket and making its measurements during a parachute descent. Capable of obtaining soundings to a much greater height than possible by balloon or aircraft.

roentgen
A unit of radiation exposure equal to 2.58×10^{-4} coulomb per kg.

Roger
'I have received all of your last transmission.' It should not be used to answer a question requiring a yes or no answer.

roll cloud
A turbulent altocumulus-type cloud formation found in the lee of some large mountain barriers. The air in the cloud rotates around an axis parallel to the mountain range. Sometimes refers to part of the cloud base along the leading edge of a cumulonimbus cloud; this is formed by rolling action in the wind shear region between cool downdrafts within the cloud and warm updrafts outside the cloud. Sometimes called rotor cloud.

rotary converter
For converting a.c. to d.c., an a.c. motor drives a d.c. generator.

rotor cloud   See ROLL CLOUD.

rotorcraft
A heavier-than-air aircraft that depends principally for its support in flight on the lift generated by one or more rotors.

rotorcraft-load combination
The combination of a rotorcraft and an external load, including the external load attaching means. Rotorcraft-load combinations are designated as Class A, Class B and Class C, as follows:

(1) Class A rotorcraft-load combination. The external load cannot move freely, cannot be jettisoned and does not extend below the landing gear.
(2) Class B rotorcraft-load combination. The external load is jettisonable and is lifted free of land or water during the rotorcraft operation.
(3) Class C rotorcraft-load combination. The external load is jettisonable and remains in contact with land or water during the rotorcraft operation.

route
A defined path, consisting of one or more tracks in a horizontal plane, which aircraft traverse over the surface of the earth. See also AIRWAY, JET ROUTE, PUBLISHED ROUTE, and UNPUBLISHED ROUTE.

route segment
As used in air traffic control, a part of a route that can be defined by two navigational fixes, two navaids, or a fix and a navaid. See also FIX and ROUTE.

route segment [ICAO]
A portion of a route to be flown, as defined by two consecutive significant points specified in a flight plan.

route stage
A route or portion of a route flown without an intermediate landing.

Royal Aeronautical Society
Formed to promote the advancement of aeronautical science in all its branches. The society has divisions in Australia, New Zealand and Africa. It also has a number of branches in the United Kingdom and Ireland. Founded under the title of the Aeronautical Society of Great Britain. The Institution of Aeronautical Engineers was founded in 1920 and incorporated into the Royal Aeronautical Society in 1927. The Helicopter Association of Great Britain was founded in 1945 and incorporated into the Royal Aeronautical Society in 1960. The Society of Licensed Aircraft Engineers and Technologists was founded in 1943, reconstituted in 1962 and incorporated into the Society in 1987. Specialist sections: the Astronautics and Guided Flight Section, the Graduates' and Students Section and the Rotorcraft Section. In addition there are groups concerned with agricultural aviation, air law, air transport, aviation medicine, avionics, flight simulation, history, aero-marine, mechanical/ structures groups, light aeroplanes, management studies, man-powered aircraft, materials, production, propulsion and test pilots.

runway
A defined rectangular area, on a land airport, prepared for the landing and takeoff run of aircraft along its length. Runways are normally numbered in relation to their magnetic direction, rounded off to the nearest 10°, e.g. Runway 18, Runway 25. See also PARALLEL RUNWAYS.

runway condition reading [US]
Numerical decelerometer readings relayed by air traffic controllers at USAF and certain civil bases for use by the pilot in determining runway braking action. These readings are routinely relayed only to USAF and Air National Guard Aircraft. See also BRAKING ACTION.

runway crossing procedures
At an aerodrome, aircraft and vehicles which are required to cross active runways will be issued instructions by the ground movement controller, which will include a clearance limit at which the aircraft or vehicle will be required to hold short of the active runway.

runway end safety area
An area symmetrical about the extended runway centreline and adjacent to the end of the strip primarily intended to reduce the risk of damage to an aeroplane undershooting or over-running the runway.

runway gradient
The average slope, measured as a percentage, between two ends or points on a runway. Runway gradient is depicted on US government aerodrome sketches when total runway gradient exceeds 0.3%.

runway in use
Any runway, or runways, currently being used for taking off or landing. When multiple runways are used, they are all considered active runways.

runway markings

(1) Basic marking. Markings on runways used for operations under visual flight rules consisting of centreline markings and runway direction numbers and, if required, letters.

(2) Instrument marking. Markings on runways served by non-visual navigation aids and intended for landings under instrument weather conditions, consisting of basic markings plus threshold markings.

(3) All-weather (precision instrument) marking. Markings on runways served by non-visual precision approach aids and on runways having special operational requirements, consisting of instrument markings plus landing zone markings and side strips.

runway profile descent

An instrument flight rules (IFR) air traffic control arrival procedure to a runway published for pilot use in graphic and/or textual form. May be associated with a STAR. Runway profile descents provide routeing, and may depict crossing altitudes, speed restrictions and headings to be flown from the en-route structure to the point where the pilot will receive clearance for, and execute, an instrument approach procedure. A runway profile descent may apply to more than one runway if so stated on the chart.

runway strip

A defined area including the runway and stopway, if provided, intended to reduce the risk of damage to aircraft running off a runway.

runway temperature

The temperature of the air just above a runway, ideally at engine and/or wing height, used in the determination of density altitude. Useful at airports when critical values of density altitude prevail.

runway guard lights

Pairs of alternatively flashing ground-mounted yellow lights at each side of the taxiways, where they connect with a runway.

runway visibility

The meteorological visibility along an identified runway determined from a specified point on the runway, may be determined by a transmissometer, or by an observer.

runway visual range

A visibility value which indicates the horizontal distance the pilot will see centreline or edge lights or runway markings from the approach end. This information should be available when either the horizontal visibility or the RVR is less than 1500 m. The provision of such information is essential for CAT II and CAT III operations. A secondary power supply should be provided for RVR observing systems which use instrumental means.

S

safety advisory [US]

A safety advisory issued by ATC to aircraft under its control if ATC is aware that the aircraft is at an altitude that, in the controller's judgement, places the aircraft in unsafe proximity to terrain obstructions or other aircraft. The controller may discontinue the issue of further advisories if the pilot advises he is taking action to correct the situation or has the other aircraft in sight.

(1) Terrain/obstruction advisory. A safety advisory issued by ATC to aircraft under its control if ATC is aware that the aircraft is at an altitude that, in the controller's judgment, places the aircraft in unsafe proximity to terrain/obstructions, e.g. 'Low altitude alert, check your altitude immediately'.

(2) Aircraft conflict advisory. A safety advisory issued by ATC to aircraft under its control if ATC is aware of an aircraft that is not under its control at an altitude that, in the controller's judgement, places both aircraft in unsafe proximity to each other. With the alert ATC will offer the pilot an alternate course of action when feasible, e.g. 'Traffic alert, advise you turn right heading zero niner zero or climb to eight thousand immediately'.

The issue of a safety advisory is contingent upon the capability of the controller to have an awareness of an unsafe condition. The course of action provided will be predicated on other traffic under ATC control. Once the advisory is issued, it is solely the pilot's prerogative to determine what course of action, if any, he will take.

safety altitude   See AREA MINIMUM ALTITUDE.

sail back

A manoeuvre during high wind conditions (usually with power off) where float plane movement is controlled by use of the flying controls, water rudders, opening and closing doors and operation of flaps.

St Elmo's fire

A luminous brush discharge of electricity from elevated objects, such as the masts and yardarms of ships, lightning rods, steeples, etc., occurring in stormy weather. Also called corposant.

St John Ambulance Air Wing

Objectives: A UK voluntary organisation (St John Ambulance Brigade) which undertakes the rapid transportation by air of transplant donor organs, the surgical teams involved and occasionally patients for emergency transplant surgery.

Santa Ana

A hot, dry, foehn wind, generally from the north east or east, occurring west of the Sierra Nevada Mountains especially in the pass and river valley near Santa Ana, California.

satellite constellation The arrangement in space of a set of satellites.

satellites
(1) Planets which revolve around stars, or moons which revolve around planets.
(2) Man-made artificial satellites for communication around the curve of the earth, research into or from the upper atmosphere, weather observations, military intelligence, etc. The first was launched by the USSR in 1957.

saturated adiabatic lapse rate
The rate of decrease of temperature with height as saturated air is lifted with no gain or loss of heat from outside sources. Varies with temperature, being greatest at low temperatures. See also ADIABATIC PROCESS and DRY ADIABATIC LAPSE RATE.

saturated air
Air that contains the maximum amount of water vapour it can hold at a given pressure and temperature (relative humidity of 100%).

say again
Used to request a repeat of the last transmission. Usually specifies transmission or portion thereof not understood or received, e.g. 'Say again all after abeam VOR . . .'.

say altitude
Used by ATC to ascertain an aircraft's specific altitude/flight level when the aircraft is climbing or descending.

say heading
Used by ATC to request an aircraft heading. The pilot should state the actual heading of the aircraft.

S-band
A general term for frequencies in the order of 3000 MHz.

scheduled duty
The allocation of a specific flight or flights to a member of the flight crew within the rostered or planned duty period.

Scheduled Maintenance Inspection [UK]
Any group of inspections or tests called up by the Maintenance Schedule.

schlieren photography
The photography of turbulence in fast moving fluids made possible by the changes in density and refractive index of the fluid.

scud
Small detached masses of stratus fractus clouds below a layer of higher clouds, usually nimbostratus.

sea breeze
A coastal breeze blowing from sea to land, caused by the temperature difference when the land surface is warmer than the sea surface.

sea fog
A type of advection fog formed when air that has been lying over a warm surface is transported over a colder water surface.

sea lane
A designated portion of water outlined by visual surface markers for, and intended to be used by, aircraft designed to operate on water.

sea-level engine
A reciprocating aircraft engine having a rated takeoff power that is produced only at sea level. Also called a normally aspirated engine.

sea-level pressure
The atmospheric pressure at mean sea level, either directly measured by stations at sea level or empirically determined from the station pressure and temperature by stations not at sea level. Used as a common reference for analyses of surface pressure patterns.

search and rescue (SAR)
A service that seeks missing aircraft and assists those found to be in need of assistance. It is a co-operative effort using the facilities and services of available State and local agencies. Information pertinent to search and rescue should be passed through any air traffic facility or be transmitted directly to the Rescue Co-ordination Centre by telephone.

search and rescue facility
A facility responsible for maintaining and operating a search and rescue (SAR) service to render aid to persons and property in distress.

search and rescue region
An area of defined dimensions within which search and rescue service is provided.

search and rescue service unit
A generic term meaning, as the case may be, rescue co-ordination centre, rescue subcentre or alerting post.

search (DME)
The condition which exists when a DME interrogator is attempting to acquire and lock onto the response to its own interrogations from the selected transponder.

sea smoke   See STEAM FOG.

second
(1) SI unit of time, $\frac{1}{60}$ of a minute, $\frac{1}{3600}$ of an hour or the duration of 9 192 631 770 periods of radiation between two hyperfine levels of the ground state of the caesium – 133 atom.
(2) Angular, $\frac{1}{60}$ of a minute of an arc, $\frac{1}{3600}$ of a degree.

secondary area
A defined area on each side of the primary area located along the nominal flight track in which decreasing obstacle clearance is provided. See also PRIMARY AREA.

secondary frequency
The radiotelephony frequency assigned to an aircraft as a second choice for air–ground communication in a radiotelephony network.

secondary radar
A radar system which requires a co-operative target: a radio link is established by an interrogator, the return or reply being supplied by a transponder on receipt of the interrogation.

secondary surveillance radar
A system of radar using ground interrogators and airborne transponders to determine the position of aircraft in range and azimuth and, when the agreed modes and codes are used, height and identity as well.

sector
That section of a flight between an aircraft moving under its own power until it next comes to rest after landing, on the designated parking position.

sector visibility [US]
Meteorological visibility within a specified sector of the horizon circle.

secular variation
Relating to the magnetic pole. It is thought that the magnetic north pole rotates around the axial north pole in 930 years. The estimated steady variation of magnetic declination is called the secular variation.

security (aviation)
A combination of measures and human and material resources intended to safeguard civil aviation against acts of unlawful interference.

security control
A means by which the introduction of weapons or articles likely to be utilised to commit an act of unlawful interference can be prevented.

security equipment
Devices of a specialised nature for use, individually or as part of a system, in the prevention or detection of acts of unlawful interference with civil aviation and its facilities

security programme
Measures adopted to safeguard civil aviation against acts of unlawful interference.

sedative
A drug that reduces anxiety and excitement and may induce sleep.

see and avoid
A visual procedure wherein pilots of aircraft flying in visual meteorological conditions (VMC), regardless of type of flight plan, are charged with the responsibility to observe the presence of other aircraft and to manoeuvre their aircraft as required to avoid the other aircraft.

segmented circle
A system of visual indicators designed to provide traffic pattern information at airports without operating control towers.

segments of an instrument approach procedure

An instrument approach procedure may have as many as four separate segments depending on how the approach procedure is structured.

(1) Initial approach. The segment between the initial approach fix and the intermediate fix or the point where the aircraft is established on the intermediate course or final approach course.

(2) Intermediate approach. The segment between the intermediate fix or point and the final approach fix.

(3) Final approach. The segment between the final approach fix or point and the runway, airport or missed approach point.

(4) Missed approach. The segment between the missed approach point, or point of arrival at decision height, and the missed approach fix at the prescribed altitude.

segments of an instrument approach procedure [ICAO]

A series of predetermined manoeuvres by reference to flight instruments with specified protection from obstacles from the initial approach fix, or where applicable, from the beginning of a defined arrival route to a point from which a landing can be completed and thereafter, if a landing is not completed, to a position at which holding or en-route obstacle clearance criteria apply.

(1) Initial approach. That part of an instrument approach procedure consisting of the first approach to the first navigational facility associated with the procedure; or to a predetermined fix.

(2) Intermediate approach. That part of an instrument approach procedure from the first arrival at the first navigational facility, or predetermined fix, to the beginning of the final approach.

(3) Final approach. That part of an instrument approach procedure from the time the aircraft has:

(a) completed the last procedure turn or base turn where one is specified; or
(b) crossed a specified fix; or
(c) Intercepted the last track specified for the procedures, until it has crossed a point in the vicinity of an aerodrome from which: (i) a landing can be made; or (ii) a missed approach procedure is initiated.

(4) Missed approach procedure. The procedure to be followed if, after an instrument approach, a landing is not effected and occurring normally:

(a) when the aircraft has decended to the decision height and has not established visual contact; or
(b) when directed by air traffic control to pull up or to go around again.

selective call system

A system used on international frequencies to improve ground-to-air communications by providing an automatic and selective method of calling aircraft from a radiotelephony aeronautical station.

self-exciting generator

Has magnets which are activated by current drawn from the output of the generator.

semi-monocoque
A type of construction in which longitudinal and transverse members relieve the skin of some of its load.

sending unit/controller
Air traffic services unit/air traffic controller transmitting a message.

sensitivity time control
A radar circuit designed to correct for range attenuation so that echo intensity on the scope is proportional to reflectivity of the target regardless of range.

separation
In air traffic control, the spacing of aircraft to achieve their safe and orderly movement in flight, and while landing and taking off.

separation [ICAO]
Spacing between aircraft, levels or tracks.

separation minima
The minimum longitudinal, lateral or vertical distances by which aircraft are spaced through the application of air traffic control procedures

series aircraft [UK]
An aircraft which is similar in every essential respect to the design of an aircraft for which a certificate of airworthiness and, where applicable, a type certificate, have previously been issued.

serious injury [ICAO]
An injury which is sustained by a person in an accident and which:

(1) requires hospitalisation for more than 48 hours, commencing within seven days from the date the injury was received; or
(2) results in a fracture of any bone (except simple fractures of fingers, toes or nose); or
(3) involves lacerations which cause severe haemorrhage, nerve, muscle or tendon damage; or
(4) involves injury to any internal organ; or
(5) involves second or third degree burns, or any burns affecting more than five per cent of the body surface; or
(6) involves verified exposure to infectious substances or injurious radiation.

service area (world area forecast system)
A geographical area within which a regional area forecast centre is responsible for supplying area forecasts to meteorological authorities and other users.

servomechanism
Changes a small mechanical motion or force into a much larger mechanical motion or force. The output is in proportion to the input, e.g. servo assisted brakes and steering on motor vehicles.

'shadient' relief
A system of showing relief on charts by a shaded picture effect, instead of by the convential use of contours and gradient tints.

shock wave
A very narrow band of high temperature pressure and density in which air pressure changes from subsonic to supersonic. Such waves may accumulate in one region with intensified effect.

shoreline
A line following the general contour of the shore, except that in cases of inlets or bays less than 30 NM in width, the line shall pass directly across the inlet or bay to intersect the general contour on the opposite side.

short circuit
A circuit (usually accidentally formed) in which an excessively large current flows because of insufficient resistance (often the cause of fire).

short range clearance
A clearance issued to a departing IFR flight that authorises IFR flight to a specific fix short of the destination while air traffic control facilities are co-ordinating and obtaining the complete clearance.

short takeoff and landing aircraft
An aircraft that, at some weight within its approved operating weight, is capable of operating from a STOL runway in compliance with the applicable STOL characteristics, e.g. airworthiness operations, noise and pollution standards.

shoulder
An area adjacent to the edge of a paved surface so prepared as to provide a transition between the pavement and the adjacent surface for aircraft running off the pavement.

shower
Precipitation from a convective cloud. Characterised by the suddenness with which it starts and stops, by the rapid changes of intensity and, usually, by rapid changes in the appearance of the sky.

shunt
An electrical conductor connected in parallel to a device in order to reduce the current running through it.

sideband
The band of frequencies on either side of a modulated carrier wave.

side lobe suppression
The technique used in SSR to ensure that aircraft are not interrogated by signals other than those radiated by the main beam of the aerial array.

sidestep manoeuvre [US]
A visual manoeuvre accomplished by a pilot at the completion of an instrument approach to permit a straight-in landing on a parallel runway not more than 1200 ft to either side of the runway to which the instrument approach was conducted.

siemens
The SI unit of conductance. The reciprocal of the ohm, i.e. $S = 1/\Omega$ and $\Omega = 1/S$.

sievert
The unit of radiation dose equivalent corresponding to 1 joule per kg.

SIGMET information [ICAO]
Information prepared by a meteorological watch office regarding the occurrence or expected occurrence of one or more of the following phenomena:

(1) At subsonic cruising levels: active thunderstorm area, tropical revolving storm, severe line squall, heavy hail, severe turbulence, severe icing, marked mountain waves, widespread sandstorm/duststorm.
(2) At transonic levels and supersonic cruising levels: moderate or severe turbulence, cumulonimbus clouds, hail.

SIGMET significant meteorological information
A weather advisory issued concerning weather significant to the safety of all aircraft. SIGMET advisories cover severe and extreme turbulence, severe icing and widespread dust or sandstorms that reduce visibility to less than three miles.

signal area
An area on an aerodrome used for the display of ground signals.

signal reliability
The probability that a signal-in-space of specified characteristics is available to an aircraft.
Note: This definition refers to the probability that the signal is present for the specified period of time.

significant obstacle (aerodromes)
Any natural terrain feature or man-made fixed object, permanent or temporary, which has vertical significance in relation to adjacent and surrounding features and which is considered a potential hazard to the safe passage of aircraft in the type of operation for which the individual chart series is designed.

significant point
A specified geographical location used in defining an ATS route or the flight path of an aircraft and for other navigation and ATS purposes.

silent radar transfer
Transfer of radar control without verbal exchange.

simplex
A method in which telecommunication between two stations takes place in one direction at a time.
Note: In application to the aeronautical mobile service this method may be subdivided as follows:

(1) single channel simplex;
(2) double channel simplex;
(3) offset frequency simplex.

simplified directional facility [US]
A navaid used for non-precision instrument approaches. The final approach track is similar to that of an ILS localiser except that the simplified directional facility track may be offset from the runway, generally not more than 3°.

simulated flameout
A practice approach by a jet aircraft (normally military) at idle thrust to a runway. The approach may start at a relatively high altitude over a runway (high key) and may continue on a relatively high and wide downwind leg with a high rate of descent and a continuous turn to final. It terminates in a landing or low approach. The purpose of this approach is to simulate a flameout.

simultaneous ILS approaches
An approach system permitting simultaneous ILS approaches to airports having parallel runways. Integral parts of a total system and ILS, radar, communications, ATC procedures and appropriate airborne equipment.

single channel simplex [RTF]
Simplex using the same frequency channel in each direction.

single direction routes [US]
Preferred IFR routes which are sometimes depicted on high altitude en-route charts and which are normally flown in one direction only.

single frequency approach [US]
A service provided under a letter of agreement to military single-piloted turbojet aircraft that permits use of a single UHF frequency during approach for landing. Pilots will not normally be required to change frequency from the beginning of the approach to touchdown except that pilots conducting an en-route descent are required to change frequency when control is transferred from the air route traffic control centre to the terminal facility.

single piloted aircraft
A military turbojet aircraft possessing one set of flight controls, tandem cockpits or two sets of flight controls but operated by one pilot is considered single piloted by ATC when determining the appropriate air traffic service to be applied.

skin-friction
Resistance to motion due to friction between the air and the surface of a solid body moving through it.

skip distance
The minimum distance at which a skywave can be received. Sky waves transmitted vertically, or nearly so, are not reflected but pass through the ionosphere into space.

slant course line [ILS]
The line formed at the intersection of the course surface and the plane of the nominal ILS glide path.

slant range
The line-of-sight distance between two points not at the same elevation.

slant visibility
For an airborne observer, the distance at which he can see and distinguish objects on the ground.

slash
A radar beacon reply displayed as an elongated target.

sleet
Generally transparent, globular, solid grains of ice which have formed from the freezing of raindrops, or the refreezing of largely melted snowflakes when falling through a below-freezing layer of air near the earth's surface.

slow switching channel
A sequencing global positioning system receiver channel that switches too slowly to allow the continuous recovery of the data message.

slow taxy
To taxy a floatplane at low power or low rpm.

slug
A unit of mass equal to 32.174 lb.

slush
Water-saturated snow which with a heel-and-toe slap-down motion against the ground will be displaced with a splatter: specific gravity 0.5 up to 0.8.
Note: Combinations of ice, snow and/or standing water may, especially when rain, rain and snow, or snow is falling, produce substances with specific gravities in excess of 0.8. These substances, due to their high water/ice content, will have a transparent rather than a cloudy appearance and, at the higher specific gravities, will be readily distinguishable from slush.

small aircraft [UK]
For the purposes of wake turbulence categorisation, aircraft between 17 000 kg and 40 000 kg are categorised by the United Kingdom as 'small'.

small aircraft [US]
Aircraft of 12 500 lb or less, maximum certificated takeoff weight.

small circle
The intersection with the earth's surface of any plane which does not pass through the centre of the earth.

smog
Fog containing dust and carbon particles and dissolved SO_2. Found over industrial cities.

smoke
A restriction to visibility resulting from combustion. Consists mainly of fine particles of carbon.

snow
Precipitation in the form of white or translucent ice crystals, chiefly in complex branched hexagonal form and often clustered into snowflakes.

snow flurry
Popular term for snow shower, particularly of a very light and brief nature.

snow grains
Precipitation of very small, white, opaque grains of ice, similar in structure to snow crystals. The grains are fairly flat or elongated, with diameters generally less than 0.04 inch (1 mm).

snow on the ground
(1) Dry snow. Snow which can be blown if loose or, if compacted by hand, will fall apart again upon release. Specific gravity: up to but not including 0.35.
(2) Wet snow. Snow which, if compacted by hand, will stick together and tend to form a snowball. Specific gravity: 0.35 up to but not including 0.5.
(3) Compacted snow. Snow which has been compressed into a solid mass that resists further compression and will hold together or break up into lumps if picked up. Specific gravity: 0.5 and over.

snow pellets
Precipitation consisting of white, opaque, approximately round (sometimes conical), ice particles, having a snow-like structure and about 0.08 to 0.2 inch in diameter. Crisp and easily crushed, differing in this respect from snow grains. Rebound from a hard surface and often break up.

SNOWTAM
A special series NOTAM notifying the presence or removal of hazardous conditions due to snow, ice, slush or standing water associated with snow, slush and ice on the movement area by means of a specific pro forma.

sodium box
Sodium approach lights arranged in box formation.

soft hail
White, opaque, round pellets of snow.

soft iron
Contains very little carbon. Cannot be made into permanent magnets because it quickly loses it magnetism.

software
Programs, languages and procedures of a computer system. No part of software has a physical existence other than as written down on paper or stored in code as represented by the state of a signal or device.

solar cell
An electric cell which converts radiant energy from the sun into electrical energy.

solar energy
5 million HP per square mile or 2 calories per minute per square centimetre or 1400 Joules per second per square metre.

solar radiation
The total electromagnetic radiation emitted by the sun.

solo flight time
Flight time during which a student pilot is the sole occupant of an aircraft.

solstice
The dates upon which the sun reaches its greatest declination north (21 June) and south (21 December), the parallels of latitude (23½% N and S) being called, respectively, the Tropic of Cancer and the Tropic of Capricorn.

somatogravic illusion
Sometimes experienced by pilots. A rapid acceleration during takeoff can excessively stimulate the sensory organs for gravity and linear acceleration, and so creates the illusion of being in a nose-up attitude. The disoriented pilot may push the aircraft into a nose-low or dive attitude. A rapid deceleration by quick reduction of the throttle(s) can have the opposite effect, with the disoriented pilot pulling the aircraft into a nose-up or stall attitude.

sonic boom
Caused by the shock wave of an aircraft travelling at supersonic speeds.

sounding balloon
A free, unmanned balloon instrumented and/or observed for obtaining a sounding of the atmosphere.

source region
An extensive area of the earth's surface characterised by relatively uniform surface conditions, where air masses remain long enough to take on characteristic temperature and moisture properties imparted by that surface.

space segment
The part of the whole GPS system that is in space, i.e. the satellites.

spacewave
A radio wave which travels in a straight line being neither refracted nor reflected.

span
The distance between the extreme tips of wings or other surfaces when measured in respect of the aircraft's plane of symmetry.

spare parts
Articles of a repair or replacement nature for incorporation in an aircraft, including engines and propellers.

speak slower
Used in verbal communications as a request to reduce speech rate.

special emergency
A condition of air piracy or other hostile act by a person(s) aboard an aircraft that threatens the safety of the aircraft or its passengers.

Special rules area
An area within an FIR or UIR within which aircraft are required to comply with instructions of air traffic control. Special rules areas extend upwards from a specified altitude or flight level and have an upper limit.

Special rules zone

An airspace within an FIR and within which aircraft are required to comply with the instructions of air traffic control. Special rules zones extend upwards from ground level to a specified altitude or flight level.

Special use airspace [US]

Airspace of defined dimensions identified by an area on the surface of the earth wherein activities must be confined because of their nature and/or wherein limitations may be imposed upon aircraft operations that are not a part of those activities. Types of special airspace are:

(1) Alert area: Airspace that may contain a high volume of pilot training activities or an unusual type of aerial activity, neither of which is hazardous to aircraft. Alert areas are depicted on aeronautical charts for the information of non-participating pilots. All activities within an alert area are conducted in accordance with Federal Aviation Regulations and pilots of participating aircraft as well as pilots transitting the area are equally responsible for collision avoidance.
(2) Controlled firing areas. Airspace wherein activities are conducted under conditions so controlled as to eliminate hazards to non-participating aircraft and ensure the safety of persons and property on the ground.
(3) Military operations are a (MOA). A MOA is an airspace assignment of defined vertical and lateral dimensions established outside positive control areas to separate/segregate certain military activities from IFR traffic and to identify for VFR traffic where these activities are conducted.
(4) Prohibited area. Designated airspace within which the flight of aircraft is prohibited.
(5) Restricted area. Airspace designated within which the flight of aircraft, while not wholly prohibited, is subject to restriction. Most restricted areas are designated joint use and IFR/VFR operations in the area may be authorised by the controlling ATC facility when it is not being utilised by the using agency. Restricted areas are depicted on en-route charts. Where joint use is authorised the name of the ATC controlling facility is also shown.
(6) Warning area. Airspace that may contain hazards to non-participating aircraft in international airspace.

Special VFR clearance

A special VFR clearance may be requested without submission of a flight plan. Brief details of the proposed flight should be passed to the appropriate air traffic control unit.

Special VFR conditions

Weather conditions in a control zone that are less than basic VFR and in which some aircraft are permitted flight under visual flight rules.

Special VFR flight

A controlled VFR flight authorised by air traffic control to operate within a control zone under meteorological conditions which are below the visual meteorological conditions.

Special VFR flight [ICAO]
A controlled VFR flight authorised by air traffic control to operate within a control zone under meteorological conditions which are below the visual meteorological conditions.

Special VFR flight [UK]
In the United Kingdom, a flight in a control zone or in special rules airspace where provision is made for such flights, in respect of which the appropriate air traffic control unit has given permission for the flight to be made in accordance with special instructions given by that unit, instead of in accordance with the instrument flight rules.

Special VFR operations
Aircraft operating in accordance with clearances within control zones in weather conditions less than the basic VFR weather minima. Such operations must be requested by the pilot and approved by ATC.

specific gravity   See RELATIVE DENSITY.

specific heat
The heat lost or gained when a substance changes temperature without changing its state (phase). It is expressed as joules per kg per °C, or calories per grams per °C, or British thermal units per pound per °F.

specific humidity
The ratio by weight of water vapour in a sample of air to the combined weight of water vapour and dry air. See also MIXING RATIO.

speed
Distance divided by time. Speed in a specified direction is called 'velocity'.

speed adjustment
An ATC procedure used to request pilots to adjust aircraft speed to a specific value for the purpose of providing desired spacing. Examples of speed adjustments are:

(1) 'Increase/reduced speed to Mach point (number)'.
(2) 'Increase/reduce speed to (speed in knots)', or 'Increase/reduce speed (number of) knots'.
(3) 'If practical, reduce speed to (speed)', or 'If practical, reduce speed (number of) knots'.

speed brakes
Moveable aerodynamic devices on aircraft that reduce airspeed during descent and landing.

spread spectrum
A system in which the transmitted signal is spread over a frequency band much wider than the minimum bandwidth needed to transmit the information being sent. This is done by modulating with a pseudo-random code, for global positioning systems.

squall
A sudden increase in wind speed by at least 15 knots to a peak of 20 knots or more and lasting for at least one minute. Essential difference between a gust and a squall is the duration of the peak speed.

squall line
Any non-frontal line or narrow band of active thunderstorms (with or without squalls); a mature instability line.

squawk
Activate specific modes/codes/functions on the transponder, e.g. 'Squawk three/ alpha, two one zero five, low'.

SSR response   See RADAR RESPONSE.

stability
A state of the atmosphere in which the vertical distribution of temperature is such that a parcel of air will resist displacement from its initial level.

stack departure time
The time at which an aircraft is required to leave the holding facility to commence its approach.

stainless steels
Steels containing 12%–20% chromium.

stall
The loss of lift caused by disruption and breakdown of the airflow over a wing surface.

stalloy
Steel containing 3½% silicon. Used in a.c. electrical apparatus with rapidly alternating magnetic fields.

Standard [ICAO]
Any specification for physical characteristics, configuration, material, performance, personnel or procedure, the uniform application of which is recognised as necessary for the safety or regularity of international air navigation and to which ICAO Contracting States will conform in accordance with the Convention. In the event of impossibility of compliance, notification to the Council is compulsory under Article 38 of the Convention.

standard atmosphere
When the term 'standard atmosphere' is used in any standards for the airworthiness of aircraft that are applicable to aircraft the prototype of which is submitted to the appropriate national authorities for certification, it means an atmosphere defined as follows:

(1) the air is a perfect dry gas;
(2) the physical constants are:
    Sea level mean molecular mass:
    $M_0 = 28.9644 \times 10^{-3}\,\text{kg/mol}$

Sea level atmospheric pressure:
p_0 = 1013.250 hPa (1013.250 mb)
Sea level temperature
t_0 = 15°C
T_0 = 288 15K
Sea level atmospheric density:
p_0 = 12250 kg/m³
Temperature of the ice point:
T_i = 273.15 K
Universal gas constant:
R^* = 8.31432 (J/mol)/K

(3) the temperature gradient from 5000 standard geopotential metres below sea level to an altitude at which the air temperature becomes −56.5°C is − 0.0065°C per standard geopotential metre; from that level (11 000 standard geopotential metres) to an altitude of 20 000 standard geopotential metres the temperature gradient is (0) and from 20 000 to 32 000 standard geopotential metres the temperature gradient is +0.0010°C per standard geopotential metre.

standard DME arrival
Designated arrival routes based upon DME information and charted to expedite clearances for final approaches.

standard instrument departure
A pre-planned instrument flight rule (IFR) air traffic control departure printed for pilot use in graphic and/or textual form. SIDs provide transition from the terminal to the appropriate en-route structure.

standard isobaric surface
An isobaric surface used on a world-wide basis for representing and analysing the conditions in the atmosphere.

standard parallel
One of the two parallels of latitude at which the cone representing a Lambert projection band intersects the sphere; in any projection, a parallel along which the scale is true.

standard positioning service
The normal civilian positioning accuracy obtained by using the single frequency C/A code in satellite global positioning systems.

standard temperature and pressure
273.15°K and 101 325 pascals.

standard terminal arrival route
A pre-planned instrument flight rule (IFR) air traffic control arrival procedure published for pilot use in graphic and/or textual form. STARs provide transition from the en-route structure to an outer fix or an instrument approach fix/arrival waypoint in the terminal area.

standard time
The civil time as standard within a zone approximately 15° of longitude in width.

stand by
The controller or pilot must pause for a few seconds, usually to attend to other duties of a higher priority. Also means to wait, as in 'stand by for clearance'. If a delay is lengthy, the caller should re-establish contact.

standby duty
A period during which an operator places restraints on a flight crew member who would otherwise be off duty.

standing wave
A wave that remains stationary in a moving fluid. In aviation operations it is used most commonly to refer to a lee wave or mountain wave.

Statement of Conformance [US]
A certifying statement that the item detailed complies with the requirements and performance standards of the Federal Aviation Regulation Technical Standard Orders.

state of manufacture
The State(s) responsible for the certification as to the airwothiness of the prototype.

state of occurrence
The State in the territory of which an accident or incidence occurs.

state of registry
The State on whose register the aircraft is entered.

state of the operator
The State in which the operator has his principal place of business or, if he has no such place of business, his permanent residence.

states of matter
Solid, liquid and gas.

static
In general, any radio interference detectable as noise in the audio stage. At broadcast frequencies, most naturally induced static is caused by non-periodic electromagnetic radiation emitted by lightning discharges acting as huge antennae; it is also known as atmospherics. Aircraft radio communication is often hindered by precipitation static due to discharges from the aircraft surfaces following self-generated electrification.

static positioning
Used in satellite navigation systems, is a location determination when the receiver's antenna is presumed to be stationary in the earth. This allows the use of various averaging techniques that improve accuracy by factors of over 1000.

statics
The mechanical study of bodies with forces acting on them but where no motion is caused.

stationary reservations [US]
Altitude reservations that encompass activities in a fixed area. Stationary reservations may include activities such as special tests of weapons systems or

equipment, certain US Navy carrier fleet and anti-submarine operations, rocket, missile and drone operations and certain aerial refuelling or similar operations.

station pressure
The actual atmospheric pressure at the observing station.

Statistics of Accidents in General Aviation
Drawn up by an ECAC working group to improve the safety of general aviation.

stator
The fixed part of an electric motor or generator which holds the stationary magnets.

statute mile
The ordinary unit of 5280 ft for measuring distance on land. It is approximately equal to 0.57 nautical miles.

steam fog
Fog formed when cold air moves over relatively warm water or wet ground.

steels
Irons containing 0.1% to 1.5% iron carbide (Fe_3C), their different properties resulting from different proportions of carbon and different methods of preparation.

steep oblique
An oblique taken with the camera depressed 10° below horizontal. No skyline visibility.

stepdown fix
A fix permitting additional descent within a segment of an instrument approach procedure by identifying a point at which a controlling obstacle has been safely overflown.

step taxy
To taxy a floatplane at full power or high rpm.

step turn
A manoeuvre used to put a floatplane in a planing configuration prior to entering an active sea lane for takeoff.

steradian
A solid angle at the centre of a sphere which subtends an area of a radius squared on the surface.

stere
A cubic metre.

stereographic
A projection used for some aeronautical charts chiefly for charts of the polar areas, where it possesses the same properties as the Lambert projection in lower latitudes.

stereo pair
Any overlapping pair of vertical photographs, usually 60°.

stero route
A routinely used route of flight established by users and ARTCCS, identified by a coded name, e.g. ALPHA 2. These routes minimise flight plan handling and communications.

stop altitude squawk
Used by ATC to inform an aircraft to turn off the automatic altitude reporting feature of its transponder. It is issued when the verbally reported altitude varies 300 ft or more from the automatic altitude report. See also ALTITUDE READOUT and TRANSPONDER.

stop and go
A procedure wherein an aircraft will land, make a complete stop on the runway and then commence a takeoff from that point. See also LOW APPROACH and OPTION APPROACH.

stop for non-traffic purposes
A landing for any purpose other than taking on or discharging passengers, cargo or mail.

stop over flight plan
A flight plan that includes two or more separate en-route flight segments with a stopover at one or more intermediate airports.

stop squawk (mode or code)
Used by ATC to tell the pilot to turn specified functions of the aircraft transponder off. See also STOP ALTITUDE SQUAWK, TRANSPONDER.

stopway
A defined rectangular area at the end of the takeoff run available, prepared and designated as a suitable area in which an aircraft can be stopped in the case of a discontinued takeoff.

stores
Articles of a readily consumable nature for use or sale on board an aircraft during flight, including commissary supplies.

storm
A marked disturbance in the normal state of the atmosphere. The term has various applications, according to the context. It is most often applied to a disturbance in which strong wind is the most prominent characteristic. It is also used for other types of disturbances, including thunderstorms, rainstorms, snowstorms, hailstorms, duststorms, sandstorms, magnetic storms, etc.

storm detection radar
A weather radar designed to detect hydrometeors of precipitation size. Used primarily to detect storms with large drops or hailstones as opposed to clouds and light precipitation of small drop size.

straight-in approach IFR
An instrument approach wherein final approach is begun without first having executed a procedure turn. Not necessarily completed with a straight-in landing or made to straight-in landing minima.

straight-in approach VFR
Entry into the traffic pattern by interception of the extended runway centreline (final approach course) without executing any other portion of the traffic pattern.

straight-in landing
A landing made on a runway aligned within 30° of the final approach course following completion of an instrument approach.

strain
The change in shape of a body resulting from stress.

stratiform
Descriptive of clouds of extensive horizontal development, as contrasted to vertically developed cumuliform clouds. Characteristic of stable air and, therefore, composed of small water droplets.

stratocumulus
A low cloud, predominantly stratiform in grey and/or whitish patches or layers, which may or may not merge. Elements are tessellated, rounded or roll-shaped with relatively flat tops.

stratosphere
The atmospheric layer above the tropopause, average altitude of base and top 7 and 22 miles respectively. Characterised by a slight average increase of temperature from base to top and is very stable. Also characterised by low moisture content and absence of clouds.

stratus
A low, grey cloud layer or sheet with a fairly uniform base. Sometimes appears in ragged patches. Seldom produces precipitation but may produce drizzle or snow grains. A stratiform cloud.

streamline
In meteorology, a line whose tangent is the wind direction at any point along the line. A flowline.

stress
Force per unit area.

strip
An area of specified dimensions enclosing a runway and taxiway to provide for the safety of aircraft operations.

sublimation
Process by which a gas is changed to a solid or a solid to a gas without going through the liquid state.

subsidence
An extensive sinking motion of air, most frequently occurring in polar highs. The subsiding air is warmed by compression and becomes more stable.

subsonic
A speed of less than Mach 0.7. It is generally accepted that on subsonic aircraft the drag coefficient remains constant up to about M 0.7 then it begins to rise appreciably, i.e., the commencement of the transonic region.

subsonic aeroplane
An aeroplane incapable of sustaining level flight at speeds exceeding flight Mach 1.

substitute route
A route assigned to pilots when any part of an airway or route is unusable because of navaid status.

sub-system
Any system which is associated with the air traffic control system as a provider and/or recipient of information relating to the provision of air traffic control service.

sulphation
The deposits of lead sulphate on the plates of a lead/acid battery which prevent it from being recharged.

summation principle
The principle states that the cover assigned to a layer is equal to the summation of the sky cover of the lowest layer plus the additional coverage at all successively higher layers up to and including the layer in question. Thus no layer can be assigned a sky cover less than a lower layer, and no sky cover can be greater than 1.0, e.g. ten/tenths (10/10), eight/eights (8/8).

superadiabatic lapse rate
A lapse rate greater than the dry-adiabatic lapse rate.

super-cooled water
Liquid water below freezing point which will suddenly change to ice (the temperature rising to zero) if shaken or if solid particles are added.

supersonic
A speed of more than Mach 1.2.

surface inversion
An inversion with its base at the surface often caused by cooling of the air near the surface as a result of terrestrial radiation, especially at night.

surface movement guidance and control system
For use under low visibility conditions at large airports. It can either take the form of a pilot self interpreted system or of a controller operated surveillance system, or even of a combination of both, depending on individual aerodrome conditions or operational requirements. Systems of the first type may consist of selectively operated taxiway centre line lights and a warning and stop system, where the pilot and the controller are alerted by audible and visual signals. At large airports there is also a requirement for special ground signs indicating to the pilot the actual position of the aircraft on the taxiway system. The second type of system comprises Surface Movement Radar (SMR) and may include, in the future other systems like SSR Mode S or ADS. Although SMR systems have been in use for a reasonable

number of years, in European airports its utilisation for aerodrome control purposes is not yet covered by ICAO provisions.

surface visibility [US]
Visibility observed from eye-level above the ground.

surveillance approach (ASR)
An instrument approach wherein the air traffic controller issues instructions, for pilot compliance, based on aircraft position in relation to the final approach course (azimuth), and the distance (range) from the end of the runway as displayed on the controller's radar scope. This controller will provide recommended altitudes on final approach if requested by the pilot.

surveillance radar
Radar equipment used to determine the position of an aircraft in range and azimuth.

sweepback
The angular set back of the main planes relative to the fuselage or hull.

sydac
5000 MH_z installation.

synchronous orbit
A 24-hour orbit enabling a satellite to remain overhead one part of the earth's surface, appearing to be stationary. At the equator the satellite's altitude would need to be 35 700 km at 3200 m/s.

synoptic chart
A weather chart describing the state of the atmosphere over a large area at a given moment.

synoptic meteorology
The branch of meteorology that deals with the analysis of weather observations made simultaneously at a number of points in the atmosphere (at the ground or aloft) over the whole or a part of the earth, and the application of the analysis to weather forecasting and other problems.

synthetic display
A display of computer-generated information, normally comprising aircraft positions and associated data presented in alphanumeric or symbolic form.

synthetic flight trainer
Any one of the following three types of apparatus in which flight conditions are simulated on the ground:

(1) A flight simulator which provides an accurate representation of the flight deck of a particular aircraft type to the extent that the mechanical, electrical, electronic, etc. aircraft systems control functions, the normal environment of flight crew members, and the performance and flight characteristics of that type of aircraft are realistically simulated.

(2) A flight procedures trainer, which provides a realistic flight deck environment,

and which simulates instrument responses, simple control functions of mechanical, electrical, electronic, etc. aircraft systems, and the performance and flight characteristics of aircraft of a particular class.

(3) A basic instrument flight trainer, which is equipped with appropriate instruments, and which simulates the flight deck environment of an aircraft in flight in instrument flight conditions.

T

tabular display
A display of information in the form of a table.

TACAN (tactical air navigation)
A method of navigation developed by the armed forces. Similar in use to the VOR plus DME: an ultra-high frequency electronic rho-theta air navigation aid that provides suitably equipped aircraft with a continuous indication of bearing and distance to the TACAN station.

TACAN only aircraft
An aircraft, normally military, possessing TACAN with DME but no VOR navigational system capability. Clearances must specify TACAN or VORTAC fixes and approaches.

tachometer
An instrument for measuring the RPM of a revolving shaft.

takeoff phase (powerplant)
The operating phase defined by the time during which the engine is operated at the rated output.

takeoff power
(1) With respect to reciprocating engines, the brake horsepower that is developed under standard, sea-level conditions, and under the maximum conditions of crankshaft rotational speed and engine manifold pressure approved for the normal takeoff, and limited in continuous use to the period of time shown in the approved engine specification.
(2) With respect to turbine engines, the brake horsepower that is developed under static conditions at a specified altitude and atmospheric temperature, and under the maximum conditions of rotorshaft rotational speed and gas temperature approved for the normal takeoff, and limited in continuous use to the period of time shown in the approved engine specification.

takeoff surface
That part of the surface of an aerodrome which the aerodrome authority has declared available for the normal ground or water run of aircraft taking-off in a particular direction.

takeoff thrust
With respect to turbine engines, the jet thrust (normally measured in lb/kg) that is developed under static conditions at a specific altitude and atmospheric tempera-ture under the maximum conditions of rotorshaft rotational speed and gas

temperature approved for the normal takeoff, and limited in continuous use to the period of time shown in the approved engine specification.

target
The indication shown on a radar display resulting from a primary radar return or a radar beacon reply.

target (ICAO)
In radar:

(1) Generally, any discrete object that reflects or retransmits energy back to the radar equipment.
(2) Specifically, an object of radar search or surveillance.

target symbol
A computer-generated indication shown on a radar display resulting from a primary radar return or a radar beacon reply.

taxi-channel
A defined path, on a water aerodrome, intended for the use of taxying aircraft.

taxi-channel lights
Aeronautical ground lights arranged along a taxi-channel to indicate the route to be followed by taxying aircraft.

taxiway
A defined path on a land aerodrome established for the taxying of aircraft and intended to provide a link between one part of the aerodrome and another, including:

(1) Aircraft stand taxi lane. A portion of an apron designated as a taxiway and intended to provide access to aircraft stands only.
(2) Apron taxiway. A portion of a taxiway system located on an apron and intended to provide a through taxy route across an apron.
(3) Rapid exit taxiway. A taxiway connected to a runway at an acute angle and designed to allow landing aeroplanes to turn off at higher speeds than are achieved on other taxiways, and thereby minimising runway occupancy times.

taxiway lights
Aeronautical ground lights arranged along a taxiway to indicate the route to be followed by taxying aircraft.

taxiway strip
An area including a taxiway intended to protect an aircraft operating on the taxiway and to reduce the risk of damage to an aircraft accidentally running off the taxiway.

taxy
The movement of an aircraft under its own power on the surface of an airport. Also describes the surface movement of helicopters equipped with wheels.

taxy/ground idle
The operating phases involving taxy and idle between the initial starting of the propulsion engine(s) and the initiation of the takeoff roll and between the time of runway turnoff and final shutdown of all propulsion engine(s).

taxy-holding position
A designated position at which taxying aircraft and other vehicles may be required to hold in order to provide adequate clearance from a runway.

taxying
Movement of an aircraft on the surface of an aerodrome under its own power, excluding takeoff and landing and, in the case of helicopters, operation over the surface of an aerodrome within a height band associated with ground effect and at speeds associated with taxying, i.e. air-taxying.

taxy into position and hold
Used by ATC to inform a pilot to taxy onto the departure runway and wait in the takeoff position. It is not authorisation for takeoff. It is used when takeoff clearance cannot immediately be issued because of traffic or other reasons.

taxy patterns
Patterns established to illustrate the desired flow of ground traffic for the different runway or airport areas available for use.

telecommunication
Any transmission, emission or reception of signs, signals, writings, images and sounds or intelligence of any nature by wire, radio, optical or other electromagnetic systems.

temporary visitor
Any person, without distinction as to race, sex, language or religion who disembarks and enters the territory of an ICAO Contracting State other than that in which that person normally resides, and remains there for not more than three months for legitimate non-immigrant purposes.

tensile strength
The pulling (tensile) stress required to break a material. Measured as force divided by area.

terminal
(1) The device by which a piece of electrical equipment is connected with the power supply.
(2) A building on an airport which links airside and landside and through which passengers pass during embarking and disembarking from aircraft.
(3) The input or ouput device of a computer.

terminal area sequencing
The process of organising traffic entering and departing from a terminal area into an orderly flow.

terminal control area
Airspace designated in the vicinity of certain major airports within which all aircraft are subject to air traffic control. Sometimes called Terminal Manoeuvring Area (TMA) (ICAO). A control area normally established at the confluence of ATS routes in the vicinity of one or more major aerodromes.

terminal control area [ICAO]
A control area normally established at the confluence of ATS routes in the vicinity of one or more major aerodromes.

terminal radar programme [US]
A national programme instituted to extend the terminal radar services. Pilot participation in the programme is urged but is not mandatory. The progressive stages of the programme are referred to as Stage I, Stage II and Stage III.

(1) Stage I/radar advisory service for VFR aircraft. Provides traffic information and limited vectoring to VFR aircraft on a workload permitting basis.
(2) Stage II/radar advisory and sequencing for VFR aircraft. Provides, in addition to Stage I service, vectoring and sequencing on a full-time basis to arriving VFR aircraft. The purpose is to adjust the flow of arriving IFR and VFR aircraft into the traffic pattern in a safe and orderly manner and to provide traffic advisory to departing VFR aircraft.
(3) Stage III/rader sequencing and separation service for VFR aircraft. Provides, in addition to Stage II services, separation between all participating aircraft. The purpose is to provide separation between all participating VFR aircraft and all IFR aircraft operating within the airspace defined as a terminal radar service area (TRSA) or terminal control area (TCA).

terminal radar service area
Airspace surrounding designated airports wherein ATC provides radar vectoring, sequencing and separation on a full-time basis for all IFR and participating VFR aircraft.

terminal route
A track defined by course, minimum altitude and distance, for bringing an aircraft from an initial approach fix to a fix from which a let-down procedure can be made. In some countries, the terminal route altitudes will be provided by ATC when not charted.

terrain following
The flight of a military aircraft maintaining a constant height above the terrain or the highest obstruction. The height of the aircraft will constantly change with the varying terrain and/or obstruction.

terrestrial guidance
Rocket or missile guidance systems which work by reference to the earth's magnetic field or its gravitational field.

terrestrial radiation
The total infra-red radiation emitted by the earth and its atmosphere.

tesla
Unit of flux density, one weber of magnetic flux per square metre.

test signal, (VOT)
A test signal for a pilot check of the accuracy of a VOR receiver. Check points for this purpose are available on the ground at many airports; others have been designated where accuracy may be tested in flight.

tetrahedron
A device normally located on uncontrolled airports in many countries and used as a landing direction indicator. The small end of a tetrahedron points in the direction of landing. At controlled airports, the tetrahedron, if installed, should be disregarded because tower instructions supersede the indicator.

that is correct
The understanding you have is right.

therm
100 000 BTU, 25 200 000 calories or 105 506 000 joules.

thermal equilibrium
The state of a system when there is no transfer of heat from one part to another.

thermal neutron analysis
A method of detecting explosives. It is a system which emits a continuous flow of low-energy neutrons and measures the gamma rays produced by the interaction of the neutrons and chemicals within explosives. Thermal neutron activation determines the composition of materials in a package by measuring the level of neutron absorption and the release of gamma rays. The system can be programmed to identify specific combinations of elements that indicate an explosive, both by type and mass. A central element in its search is nitrogen which is present in explosives.

thermobarograph
Instrument for measuring both temperature and pressure continuously.

thermocouple
A thermometer made of two wires made of two different metals joined at each end. One end is placed where the temperature must be measured and the resulting emt is measured at the other end.

thermometer
An instrument for measuring temperature, consisting of a graduated glass tube with a bulb at one end, containing alcohol or mercury.

thermosphere
The region in the upper atmosphere in which the temperature increases with altitude.

thermostat
A device for switching heating and cooling equipment on and off so as to maintain a constant temperature.

threshold
The beginning of that portion of the runway usable for landing.

threshold crossing height
The height of the glide slope above the runway threshold.

threshold lights
Aeronautical ground lights so placed as to indicate the longitudinal limits of that portion of a runway, channel or landing path usable for landing.

thrust
The force produced by a propeller, jet or rocket.

thunder
The sound emitted by rapidly expanding gases along the channel of a lightning discharge.

thunderstorm
A storm, invariably produced by a cumulonimbus cloud, and always accompanied by lightning and thunder. Usually attended by strong wind gusts, heavy rain and sometimes hail. It is usually of short duration, seldom over two hours for any one storm.

time group
Four digits representing the hour and minutes from the 24-hour clock. Time group without time zone indicators are understood to be GMT (Greenwich Mean Time), e.g. '0205'. A time zone designator is used to indicate local time, e.g. '0205M'. The end and beginning of the day are shown by '2400' and '0000' respectively. In ICAO terminology GMT has now been replaced by 'universal co-ordinated time' (UTC).

time in service
With respect to maintenance time records, the time from the moment an aircraft leaves the surface of the earth until it touches it at the next point of landing.

time slot
A period of time allocated to an aircraft to take off or to pass over a given reference point.

to certify as airworthy
To certify that an aircraft or parts thereof comply with current airworthiness requirements after being overhauled, repaired, modified or installed.

tonne
Mass equal to 1000 kgs.

to pilot
To manipulate the flight controls of an aircraft during flight time.

torching
The burning of fuel at the end of an exhaust pipe or stack of a reciprocating aircraft engine, the result of an excessive richness in the fuel-air mixture.

tornado
A violently rotating column of air, pendant from a cumulonimbus cloud, and nearly always observable as 'funnel-shaped'. It is the most destructive of all small-scale atmospheric phenomena. Sometimes called cyclone, twister.

total estimated elapsed time
For IFR flights, the estimated time required from takeoff to arrive over that designated point, defined by reference to navigation aids, from which it is intended that an instrument approach procedure will be commenced, or, if no navigation aid is associated with the destination aerodrome, to arrive over the destination aerodrome. For VFR flights, the estimated time required from takeoff to arrive over the destination aerodrome.

touch and go
An operation by an aircraft that lands and departs on a runway without stopping or exiting the runway.

touch and go landing   See TOUCH AND GO.

touchdown
(1) The point at which an aircraft first makes contact with the landing surface.
(2) Concerning a precision radar approach (PAR), the point where the glidepath intercepts the landing surface.

touchdown [ICAO]
The point where the nominal glidepath intercepts the runway.

touchdown zone
The first 3000 ft of the runway beginning at the threshold; the area used for determination of touchdown zone elevation in the development of straight-in landing minima for instrument approaches.

touchdown zone elevation
The highest elevation in the first 3000 ft of the landing surface. This is indicated on the instrument approach procedure chart when straight-in landing minima are authorised.

tower en-route control service
The control of IFR en-route traffic within delegated airspace between two or more adjacent approach control facilities. This service is designed to expedite traffic and reduce control and pilot communication requirements.

towering cumulus
A rapidly growing cumulus in which height exceeds width.

tower to tower   See TOWER EN-ROUTE CONTROL SERVICE.

tower visibility
Prevailing visibility determined from the control tower.

trace narcotics detector
A system which uses gas chromatography to detect particles emanating from concealed caches of drugs such as cocaine, heroin and amphetamines.

track
The projection on the earth's surface of the path of an aircraft, the direction of which path at any point is usually expressed in degrees from north (true, magnetic or grid).

track angle error
The angular difference between track and desired track.

track homing
The process of following a line of position which is known to pass through an intended destination or waypoint.

trade winds
Prevailing, almost continuous, winds blowing with an easterly component from the subtropical high pressure belts towards the intertropical convergence zone; northeast in the northern hemisphere, southeast in the southern hemisphere.

traffic advisories
Advisories issued to alert a pilot to other known or observed air traffic that may be in such proximity to his aircraft's position or intended route of flight to warrant his attention. Such advisories may be based on:

(1) Visual observation.
(2) Observation of radar identified and non-identified aircraft targets on an ATC radar display, or
(3) Verbal reports from pilots or other facilities.

Controllers use the word 'traffic' followed by additional information, if known, to provide such advisories, e.g. 'traffic, two o'clock, one zero miles, southbound, eight thousand'. Traffic advisory service will be provided to the extent possible depending on higher priority duties of the controller or other limitations, e.g. radar limitations, volume of traffic, frequency congestion or controller workload. Radar/non-radar traffic advisories do not relieve the pilot of his responsibility to see and avoid other aircraft. Pilots are cautioned that there are many times when the controller is not able to give traffic advisories concerning all traffic in the aircraft's proximity; in other words, when a pilot requests or is receiving traffic advisories, he should not assume that all traffic information will be issued.

traffic information zone [US]
An airspace extending upwards from the surface of the earth to a specified upper limit within which IFR flights will be informed about VFR flights conducted in meteorological conditions less than the minima specified for conducting VFR flight in controlled airspace.

traffic in sight
Used by pilots to inform a controller that previously issued traffic is in sight.

traffic no longer a factor
Indicates that the traffic described in a previously issued traffic advisory is no longer a factor to consider for the safety of the flight.

traffic orientation
Planning and application of a routeing of air traffic between specified points, or areas of origin and specific destinations.

traffic orientation system
This system forms the basis for the routeing of air traffic on the major traffic flows during the summer peak season on the understanding that the application of the routeings contained will improve the balancing of the demand on the air navigation

system, thus permitting the air traffic services to provide the most expeditious service to air traffic during the peak periods of activity.

traffic pattern
The traffic flow that is prescribed for aircraft landing at, taxying on or taking off from an airport. The components of a typical traffic pattern are upwind leg, crosswind leg, downwind leg, baseleg and final approach.

(1) Upwind leg. A flight path parallel to the landing runway in the direction of landing.
(2) Crosswind leg. A flight path at right angles to the landing runway off its upwind end.
(3) Downwind leg. A flight path parallel to the landing runway in the direction opposite to landing. The downwind leg normally extends between the crosswind leg and the base leg.
(4) Base leg. A flight path at right angles to the landing runway off its approach end. The base leg normally extends from the downwind leg to the intersection of the extended runway centreline.
(5) Final approach. A flight path in the direction of landing along the extended runway centreline. The final approach normally extends from the base leg to the runway. An aircraft making a straight-in approach VFR is also considered to be on final approach.
(6) (ICAO) Aerodrome traffic circuit. The specified path to be flown by aircraft operating in the vicinity of an aerodrome.

trajectory
The path traced out by a small volume of air in its movement over the earth's surface. The path of a projectile.

transcribed weather broadcast [US]
A continuous recording of meteorological and aeronautical information that is broadcast on L/MF and VOR facilities for pilots.

transfer of control
That action whereby the responsibility for the separation of an aircraft is transferred from one controller to another.

transfer to control [ICAO]
Transfer of responsibility for providing air traffic control service.

transfer of control point
A defined point location along the flight path of an aircraft, at which the responsibility for providing air traffic control service to the aircraft is transferred from one control unit or control position to the next.

transferring controller/facility
A controller/facility transferring control of an aircraft to another controller/facility.

transferring unit/controller
Air traffic control unit/air traffic controller in the process of transferring the

responsibility for providing air traffic control service to an aircraft to the next air traffic control unit/air traffic controller along the route of flight.

transition

(1) The general term that describes the change from one phase of flight or flight condition to another, e.g. transition from en-route flight to the approach or transition from instrument flight to visual flight.

(2) A published route (SID transition) used to connect the basic SID to one of several en-route airways/jet routes, or a published procedure (STAR transition) used to connect one of several en-route airways/jet routes to the basic STAR.

transition altitude (QNH)

The altitude in the vicinity of an airport at or below which the vertical position of an aircraft is controlled by reference to altitudes above mean sea level (amsl).

transition height (QFE)

The height in the vicinity of an airport at or below which the vertical position of an aircraft is expressed in height above the airport reference datum.

transition layer

The airspace between the transition altitude and the transition level. Aircraft descending through the transition layer will use altimeters set to local station pressure, while departing aircraft climbing through the layer will be using standard altimeter setting (QNE) of 29.92 inches of mercury or 1013.2 millibars.

transition level (QNE)

The lowest flight level available for use above the transition altitude.

transition zone

The relatively narrow region occupied by a front. The meteorological properties in this zone exhibit large variations over a short distance and possess values intermediate between the characteristics of the air masses on either side of the zone.

transmission rate

The average number of pulse pairs transmitted from a transponder per second.

transmissometer

An instrument system which shows the transmissivity of light through the atmosphere. Transmissivity may be translated either automatically or manually into visibility and/or runway visual range.

transmitting blind   See BLIND TRANSMISSION.

transponder

The airborne radar beacon receiver/transmitter portion of the air traffic control radar beacon systems (ATCRBS), that automatically receives radio signals from interrogators on the ground, and selectively replies with a specific reply pulse or pulse group only to those interrogations being received on the mode to which it is set to respond.

transponder [ICAO]
A receiver/transmitter that will generate a reply signal upon proper interrogation, the interrogation and reply being on different frequencies.

tributary station
An aeronautical fixed station that may receive or transmit messages and/or digital data, but which does not relay except for the purpose of serving similar stations connected through it to a communication centre.

tropical air
An air mass with characteristics developed over low latitudes. Maritime tropical air (mT), the principal type, is produced over the tropical and subtropical seas and is very warm and humid. Continental tropical air (cT) is produced over subtropical arid regions and is hot and very dry. Compare polar air.

tropical cyclone
A general term for a cyclone that originates over tropical oceans. By international agreement tropical cyclones have been classified according to their intensity, as follows:

(1) Tropical depression. Winds up to 34 knots (64 km/h).
(2) Tropical storm. Winds of 35 to 64 knots (65 to 119 km/h).
(3) Hurricane or typhoon. Winds of 65 knots or higher (120 km/h).

tropical disturbance [US]
The name used by the Weather Bureau for a cyclonic wind system of the tropics that is not known to have sufficient force to justify the use of the words 'storm' or 'hurricane'.

tropopause
The boundary between the troposphere and stratosphere, usually characterised by an abrupt change of lapse rate.

troposphere
That portion of the atmosphere from the earth's surface to the tropopause, i.e. the lowest 10 to 20 kilometres of the atmosphere. The troposphere is characterised by decreasing temperature with height, and by appreciable water vapour.

trough
In meteorology, an elongated area of relatively low atmospheric pressure usually associated with, and most clearly identified as, an area of maximum cyclonic curvature of the wind flow (isobars, contours, or streamlines). Compare with ridge. Also called trough line.

true altitude
The exact distance above mean sea level.

turbojet aircraft
An aircraft with a jet engine in which the energy of the jet operates a turbine that in turn operates the air compressor.

turboprop aircraft
An aircraft with a turbine engine or engines in which the energy derived from the engine gas flow operates a turbine that drives a propeller and compressor.

turbulence
Irregular motion of the atmosphere produced when air flows over a comparatively uneven surface, such as the surface of the earth, or when two currents of air flow past or over each other in different directions or at different speeds.

T-VOR (terminal very high frequency omnidirectional range station)
A very high frequency terminal omnirange station located on or near an airport and used as an approach aid.

twilight
The intervals of incomplete darkness following sunset and preceding sunrise. The time at which evening twilight ends or morning twilight begins is determined by arbitrary convention, and several kinds of twilight have been defined and used, most commonly civil, nautical, and astronomical twilight.

(1) Civil twilight. The period of time before sunrise and after sunset when the sun is not more than 6° below the horizon.
(2) Nautical twilight. The period of time before sunrise and after sunset when the sun is not more than 12° below the horizon.
(3) Astronomical twilight. The period of time before sunrise and after sunset when the sun is not more than 18° below the horizon.

twister
In the United States, a colloquial term for tornado.

two frequency glidepath system
An ILS glidepath in which coverage is achieved by the use of two independent radiation field patterns spaced on separate carrier frequencies within the particular glidepath channel.

two frequency localiser system
A localiser system in which coverage is achieved by the use of two independent radiation field patterns spaced on separate carrier frequencies within the particular localiser VHF channel.

two-way radio communications failure   See LOST COMMUNICATIONS.

type approval
A term used to describe the procedures of, or the achievement of, type certification.

type certificate
A certifying statement issued by airworthiness authorities, that the aircraft and variants referred to on the related type certificate data sheet have been approved.

type certificate data sheet
Part of the aircraft type certificate.

type record
A summary of aircraft design.

typhoon
A tropical cyclone in the eastern hemisphere with winds in excess of 65 knots (120 km/h).

U

udometer
Pluviometer, rain guage.

ultimate load
The limit load multiplied by the appropriate factor of safety.

ultra-high frequency
The frequency band between 300 and 3000 MHz. The bank of radio frequencies used for military air–ground voice communications. In some instances this may go as low as 225 MHz and still be referred to as UHF.

unable
Indicates inability to comply with a specific instruction, request or clearance.

uncertainty phase
A situation wherein uncertainty exists as to the safety of an aircraft and its occupants.

uncontrolled airspace
That portion of the airspace that has not been designated as a continental control area, control area, control zone, terminal control area or transition area, and within which ATC has neither the authority nor the responsibility for exercising control over air traffic.

undercast [US]
A cloud layer of ten-tenths (1.0) coverage (to the nearest tenth) as viewed from an observation point above the layer.

under command
An aeroplane on the surface of the water is 'under command' when it is able to execute manoeuvres as required by the International Regulations for Preventing Collisions at Sea for the purpose of avoiding other vessels.

under the hood
Indicates that the pilot is using a hood to restrict visibility outside the cockpit while simulating instrument flight. An appropriately rated pilot is required in the other control seat while this operation is being conducted.

under way
An aeroplane on the surface of the water is 'under way' when it is not aground or moored to the ground or to any fixed object on the land or in the water.

UNICOM [US]
A non-government air–ground communication facility which may provide airport advisory service at certain airports. Locations and frequencies of UNICOM are shown on aeronautical charts and publications.

United Nations Development Programme
A major financing partner for ICAO's technical assistance activities.

universe
Thought to be a curved four-dimensional space time continuum containing 10^9 galaxies totalling 10^{41} kg of matter.

unlimited ceiling
A clear sky or a sky cover that does not meet the criteria for a ceiling.

unlimited route concept
A concept of controlled airspace organisation which allows an operator complete freedom to choose the route to be taken by a flight from one point to another provided that the route is adequately defined in the flight plan and adhered to as accurately as circumstances permit.

unmanned free balloon
A non-power driven, unmanned, lighter-than-air aircraft in free flight.
Note: Unmanned free balloons are classified as heavy, medium or light in accordance with specifications contained in ICAO Annex 2, Appendix D.

UN number
The four-digit number assigned by the United Nations Committee of Experts on the Transport of Dangerous Goods to identify a substance or a particular group of substances.

unpublished route
A route for which no minimum altitude is published or charted for pilot use. It may include a direct route between navaids, a radial, a radar vector or a final approach course beyond the segments of an instrument approach procedure.

updraught
A localised upward current of air.

upper-air chart
A meteorological chart relating to a specified upper-air surface or layer of the atmosphere.

upper ATS route
A designated route in the special rules area within the upper airspace.

upper front
A front aloft not extending to the earth's surface.

upper side band
The sideband of an AM transmission which is higher than the carrier frequency.

upslope fog
Fog formed when air flows upward over rising terrain and is, consequently, adiabatically cooled to or below its initial dew point.

urgency
A condition of being concerned about safety and of requiring timely but not immediate assistance: a potential distress condition.

Urgency [ICAO]

A condition concerning the safety of an aircraft or other vehicle or person on board or in sight, but that does not require immediate assistance.

usability factor

The percentage of time during which the use of a runway or system of runways is not restricted because of the cross-wind component.

Note: Cross-wind component means the surface wind component at right angles to the runway centreline.

use interface

The way a global positioning system (GPS) receiver conveys information to the person using it. The controls and displays.

user segment

The part of the whole satellite global positioning system (GPS) that includes the receivers of GPS signals.

V

vacuum
A space in which there is no matter.

vane
A device that shows which way the wind blows. Also called weather vane or wind vane.

vapour
A gas below its critical temperature, i.e. a gas which is not too hot to be liquified by pressure.

vapour pressure
In meteorology, the pressure of water vapour in the atmosphere. Vapour pressure is that part of the total atmospheric pressure due to water vapour and is independent of the other atmospheric gases or vapours.

vapour trail   See CONDENSATION TRAIL.

variable load [UK]
The weight of the crew, of items such as the crew's baggage, removable units, and other equipment, the carriage of which depends upon the role for which the operator intends to use the aircraft for the particular flight.

variant [UK]
For the purposes of airworthiness requirements a variant (previously referred to as Prototype Modified) is 'an aircraft which embodies certain design features dissimilar to the Prototype' which are required to be investigated for certification purposes.

variant validity times
In meteorological forecasts refers to two pairs of digits denoting the hours of validity of the sub-period.

vector
(1) Any physical quantity that includes a direction, e.g. velocity. When a body is acted on by two vectors, the two vectors may be represented by two sides of a triangle, and their resultant by the third side.
(2) A heading issued to an aircraft to provide navigational guidance by radar.

vectoring (radar) [ICAO]
Provision of navigational guidance to aircraft in the form of specific headings, based on the use of radar.

veering
Shifting of the wind in a clockwise direction with respect to either space or time; opposite of backing. Commonly used by meteorologists to refer to an anticyclonic shift (clockwise in the northern hemisphere and anticlockwise in the southern hemisphere).

velocity
The rate of change of displacement, or the vector of speed.

velocity ratio
The distance moved by the applied force divided by the distance moved by the load. This also equals the mechanical advantage ratio because force × distance = load × distance.

venturi tube
A tube which gently narrows at the centre and broadens at the ends. The static pressure of a fluid flowing through it is less in the narrower section.

verify
Request confirmation of information, e.g. 'verify assigned altitude'.

verify specific direction of takeoff, or turns after takeoff [US]
Used by ATC to ascertain an aircraft's direction of takeoff and/or direction of turn after takeoff. It is normally used for IFR departures from an airport not having a control tower. When direct communication with the pilot is not possible, the request and information may be relayed through an FSS, dispatcher or by other means.

vertical planes
Planes perpendicular to the horizontal plane.

vertical reference gyro
A gyro to which gravity-controlled forces are applied by an erection system such as to maintain the spin axis in the vertical plane. Used to give signals proportional to pitch and roll.

vertical separation
Separation established by assignment of different altitudes or flight levels.

vertical separation [ICAO]
Separation between aircraft expressed in units of vertical distance.

vertical speed indicator
An instrument indicating the rate of ascent or descent of an aircraft, usually in feet per minute.

vertical takeoff and landing aircraft
Aircraft capable of vertical climbs and/or descents and of using very short runways or small areas for takeoff and landings. These aircraft include, but are not limited to, helicopters. See also SHORT TAKEOFF AND LANDING AIRCRAFT.

vertical visibility
The distance one can see upward into a surface based obscuration; the maximum height from which a pilot in flight can recognise the ground through a surface-based obscuration.

vertigo
Used to indicate disorientation which may include a sensation of giddiness and is usually caused by rapidly changing forces due to linear, radial and angular acceleration. It may also be caused by disordered function of the vestibular apparatus due to disease.

very high frequency
The frequency band between 30 and 300 MHz. Portions of this band, 108 to 118 MHz, are used for certain navaids; 118 to 137 MHz are used for civil air ground voice communications; Other frequencies in this band are used for purposes not related to air traffic control.

very high frequency omnidirectional range station (VOR)
A ground-based electronic navaid transmitting very high frequency navigation signals, 360 degress in azimuth, oriented from magnetic north. Used as a basis for navigation. The VOR periodically identifies itself by Morse code and may have an additional voice identification feature. Voice features may be used by ATC or FSS for transmitting instructions/information to pilots.

very high frequency omnidirectional range/tactical air navigation
A navigation aid providing VOR azimuth, TACAN azimuth and TACAN distance measuring equipment (DME) at one site.

very low frequency
The frequency band between 3 and 30 kH.

VFR flight
A flight conducted in accordance with the visual flight rules.

virga
Water or ice particles falling from a cloud, usually in wisps or streaks, and evaporating before reaching the ground.

viscosity
A fluid's internal resistance to flow.

visibility
The ability, as determined by atmospheric conditions and expressed in units of distance, to see and identify prominent unlighted objects by day and prominent lighted objects by night.
(1) Flight visibility. Average forward horizontal distance at which a pilot in flight may see and identify prominent unlighted objects by day, or prominent lighted objects by night.
(2) Ground visibility. Prevailing horizontal visibility near the earth's surface as reported by an accredited observer.
(3) Prevailing visibility. The horizontal distance at which objects of known distance

are visible over at least half of the horizon. Usually determined at the control tower and reported in kilometres, metres, or miles or fraction of a mile.

(4) Runway visibility by observer. Horizontal distance at which a light of about 25 candle power at night, or a dark object against the horizon sky in the daytime, can be seen by an observer near the end of the runway. Used in place of prevailing visibility in determining minimums for a particular runway.

(5) Runway visibility value (RVV). Visibility determined for a particular runway by transmissometer. Gradually being replaced by RVR.

(6) Runway visual range (RVR). The horizontal distance a pilot may expect to see a high intensity runway light, down the runway from the approach end, from a moving aircraft.

visual acuity
The ability to see. Is assessed by a person reading from an illuminated test card (with rows of different size letters), placed at a distance of six metres. The test is based on the assumption that the average normal eye is able to see clearly a letter whose overall size subtends an angle of 5 minutes of arc as it reaches the eye and the width of each stroke of each letter subtends an angle of 1 minute of arc at the eye. Since the standard testing distance is six metres, it follows that all vision is recorded in the 6-m notation; the numerator, or figure above the line in the fraction, indicating the test distance and the denominator, or figure below the line in the fraction, indicating the distance at which the test letter should be clearly seen. Thus the fraction $\frac{6}{6}$ means that at a distance of six metres the eye can read those letters which are intended to be read by the normal eye at a distance of six metres. The fraction $\frac{6}{8}$ means that the eye at a distance of six metres can only read those (larger) letters intended to be read by the normal eye at a distance of eight metres.

visual approach [ICAO]
An approach by an IFR flight when either part or all of an instrument approach procedure is not completed and the approach is executed in visual reference to terrain.

visual approach slope indicator
An airport lighting facility providing vertical visual approach slope guidance to aircraft during approach to landing by radiating a directional pattern of high intensity red and white focused light beams that indicate to the pilot that he is 'on path' if he sees red/white, 'above path' if white/white, and 'below path' if red/red. Some airports serving large aircraft have three-bar VASIs that provide two visual glidepaths to the same runway. There are three standard VASI systems and many non-standard systems used throughout the world. The standard 2-BAR VASI is referred to as 'VASI' and 'AVASI'; the 1-BAR with bisecting longitudinal lines of six lights is referred to as 'T-VASI' (located on both sides of the runway); the 3-BAR VASI is referred to as 'VASI (3-BAR)'; the non-standard VASI or 'VASI (non-std)'. The VASI, AVASI, T-VASI and AT-VASI are for conventional/sized aircraft with limited use for lor g-bodied jets. The VASI (3-BAR) system is a standard VASI or AVASI system with a second upwind or far bar added primarily for use of long-bodied jets, e.g. B-747. L-1011, C-5A. etc. All systems use a visual fixed glide path for descents to landing. VASI, AVASI, T-VASI and AT-VASI systems provide one visual glidepath; VASI (3-BAR) systems provide two visual glide paths.

visual approach slope indicator [ICAO]
VASI and AVASI systems light paths are normally set between 2.5° and 3.0°. Most systems use two light bars referred to as 'downwind' and 'upwind' bars. VASI (3-BAR) systems upper light path is normally set at 3.0°. The lower light path uses two light bars referred to as 'downwind' and 'middle' and the upper light path uses light bars referred to as 'middle' and 'upwind' bars. T-VASI and AT-VASI systems' normal descent angle is recommended to be 2.87°.

visual approach slope indicator [US]
VASI and AVASI systems light paths are normally set at 3.0°. Most systems use two light bars referred to as 'near' and 'far' bars. VASI (3-BAR) systems' lower light path is normally set at 3.0° and the upper light path is normally set at 3.25°. The lower light path uses the two light bars referred to as 'near' and 'middle' bars and the upper path uses light bars referred to as 'middle' and 'far' bars.

visual descent point [US]
A defined point on the final approach course of a non-precision straight-in approach procedure from which normal descent from the MDA to the runway touchdown point may be commenced, provided the required visual reference is established. This point is identified by an approved navigational aid. Visual descent points (VDPs) are not a mandatory part of non-precision approach procedures but are intended to provide additional guidance where they are implemented. A VASI will normally be installed on those runways served by a non-precision approach that incorporates a VDP.

visual display unit
A cathode-ray tube for displaying computer information in words or diagrams.

Visual Flight Operation Panel [ICAO]
A panel set up by ICAO to undertake a review of the classification of airspace and to standardise the types of flight operations permitted in each class.

visual flight rules
Rules that govern the procedures for conducting flight under visual conditions. The term 'VRF' is also used in the United States to indicate weather conditions that are equal to or greater than minimum VFR requirements. In addition, it is used by pilots and controllers to indicate type of flight plan.

visual holding
The holding of aircraft at selected, prominent, geographical fixes that can easily be recognised from the air.

visual manoeuvring (circling) area
The area in which obstacle clearance should be taken into consideration for aircraft carrying out a circling approach.

visual meteorological conditions
Meteorological conditions expressed in terms of visibility, distance from cloud and ceiling equal to, or better than, specified minima.

visual reference
A view of a section of the runway and/or the approach area and/or their visual aids which the pilot must see in sufficient time to assess whether or not a safe landing can be made from the type of approach being conducted.

visual separation
A means employed by ATC to separate aircraft in terminal areas. There are two ways to effect this separation:

(1) The tower controller sees the aircraft involved and issues instructions, as necessary, to ensure that the aircraft avoid each other.
(2) A pilot sees the other aircraft involved and upon instructions from the controller provides his own separation by manoeuvring his aircraft as necessary to avoid it. This may involve following another aircraft or keeping it in sight until it is no longer a factor.

VOLMET broadcast
Routine broadcast of meteorological information for aircraft in flight.

volt
The unit of electric potential difference and electromotive force which is the difference of electric potential between two points on a conductor carrying a constant current of one ampere, when the power dissipated between these points is equal to one watt.

VORTAC
VOR and TACAN beacons on the same geographical site, i.e. co-located, are termed collectively a VORTAC beacon.

vortex
In meteorology, any rotary flow in the atmosphere.

vortices
Circular patterns of air created by the movement of an aerofoil through the air when generating lift. As an aerofoil moves through the atmosphere in sustained flight, an area of low pressure in created above it. The air flowing from the high pressure area to the low pressure area around and about the tips of the aerofoil tends to roll up into two rapidly rotating vortices, cylindrical in shape. These vortices are the most predominant parts of aircraft wake turbulence and their rotational force is dependent upon the wing loading, gross weight and speed of the generating aircraft. The vortices from medium to heavy aircraft can be of extremely high velocity and hazardous to smaller aircraft.

vorticity
Turning of the atmosphere. Vorticity may be embedded in the total flow and not readily identified by a flow pattern.

(1) Absolute vorticity. The rotation of the earth imparts vorticity to the atmosphere. Absolute vorticity is the combined vorticity due to this rotation and vorticity due to circulation relative to the earth (relative vorticity).

(2) Negative vorticity. Vorticity caused by anticyclonic turning. It is associated with downward motion of the air.
(3) Positive vorticity. Vorticity caused by cyclonic turning. It is associated with upward motion of the air.
(4) Relative vorticity. Vorticity of the air relative to the earth, disregarding the component of vorticity resulting from the earth's rotation.

wake turbulence
Turbulence found to the rear of a solid body in motion relative to a fluid. In aviation terminology, the turbulence caused by the lifting surfaces of a moving aircraft.

wake turbulence categorisation
For the purpose of classifying the degree of wake turbulence produced by an aircraft, ICAO has categorised aircraft in accordance with their maximum permitted take-off weights. See also HEAVY AIRCRAFT, LIGHT AIRCRAFT, MEDIUM AIRCRAFT and SMALL AIRCRAFT.

wake turbulence characteristics
Wake vortices, sometimes referred to as wake turbulence, are present behind every aircraft, including helicopters when in forward flight, but are particularly severe when generated by heavy aircraft such as the wide-bodied jets. They are most hazardous to aircraft with a small wing span during the takeoff, initial climb, final approach and landing phases of flight. The characteristics of the wake vortex system generated by an aircraft in flight are determined initially by the aircraft's gross weight, wingspan, airspeed configuration and attitude. Subsequently these characteristics are altered by interactions between the vortices and the ambient atmosphere and eventually, after a time varying according to the circumstances from a few seconds to a few minutes after the passage of an aircraft, the effects of the wake become undetectable. For practical purposes, the vortex system in the wake of an aircraft may be regarded as being made up of two counter-rotating cylindrical air masses trailing aft from the aircraft. Typically the two vortices are separated by about three quarters of the aircraft's wingspan, and in still air they tend to drift slowly downwards and either level off, usually not more than 1000 ft below the flight path of the aircraft, or, on approaching the ground, move sideways from the track of the generating aircraft at a height approximately equal to half the aircraft's wingspan.

wall cloud
The well-defined bank of vertically developed clouds having a wall-like appearance which form the outer boundary of the eye of a well-developed tropical cyclone.

warm front
Any non-occluded front which moves in such a way that warmer air replaces colder air.

warm sector
The area covered by warm air at the surface and bounded by the warm front and cold front of a wave cyclone.

warning area
International airspace (as beyond the three-mile limit in coastal areas) which may contain hazards to flight.

washout
The twist built into an aircraft wing to reduce its angle of incidence towards the tip, thus creating a condition whereby the inner section of the wing stalls before the outer section.

water equivalent
The depth of water that would result from the melting of snow or ice.

waterspout
A tornado occurring over water; rarely, a lesser whirlwind over water, comparable in intensity to a dust devil over land.

water vapour
Water in an invisible gaseous form.

watt
Unit of power = J/s. 745.7 watts = 1 horse power. Electricially, watts may be calculated as volts × amps. One kilowatt hour = $3\,600\,000$ J = one commercial unit of electricity.

watt second
A unit of work or energy = 1 joule = 1 newton metre = 0.24 calories.

wave cyclone
A cyclone which forms and moves along a front. The circulation about the cyclone centre tends to produce a wave-like deformation of the front.

waypoint
In area navigation, a point on a selected direct route, defined by its direction and distance from a navaid which may be used for en-route navigation, position reporting, etc.

weather
The short-term variations of the atmosphere in terms of temperature, pressure, wind, moisture, cloudiness, precipitation and visibility.

weather advisory [US]
In aviation weather forecasting practice, an expression of hazardous weather conditions not predicted in the area forecast, as they affect the operation of air traffic and as prepared by the National Weather Service.

weathercock stability
A natural characteristic of an aeroplane to align itself about its yaw axis into the line of the prevailing airflow or wind.

weather radar
Radar specifically designed for observing weather. See also CLOUD DETECTION RADAR and STORM DETECTION RADAR.

weather vane
A wind vane.

weber
The magnetic flux which, linking a circuit of one turn, produces in it an electromotive force of one volt as it is reduced to zero at a uniform rate in one second.

weight and balance report
Produced for each prototype, variant and series aircraft exceeding 5700 kg netwa.

weighing record
The documentation associated with each weighing of an aircraft, the calculations made and resultant c.g. changes, if any, have to be retained with the aircraft records.

weight
Weight is defined as: force = mass × acceleration of gravity. Measured (internationally) in newtons. One kilogram-force (kgf) = approximately 10 newtons.

wet bulb
Contraction of either wet-bulb temperature or wet-bulb thermometer.

wet-bulb temperature
The lowest temperature that can be obtained on a wet-bulb thermometer in any given sample of air, by evaporation of water (or ice) from the muslin wick. Used in computing dew point and relative humidity.

wet-bulb thermometer
A thermometer with a muslin-covered bulb used to measure wet-bulb temperature.

whirl mode
A type of flutter instability involving a gyroscopic wobbling motion of the propellor nacelle powerplant system surrounding a flexible engine mount

whirlwind
A small, rotating column of air. May be visible as a dust devil.

wilco
I have received your message, understand it and will comply with it.

willy-willy
A tropical cyclone of hurricane strength near Australia.

wind
Air in motion relative to the surface of the earth. Generally used to denote horizontal movement.

wind angle
The angle between the true track and (course or heading, as the case may be) and the direction from which the wind is blowing. Measured from the true track course or heading towards the right or left from 000° up to 180°.

wind correction angle
The angle (relative to the true track, course or heading) at which an aircraft must be headed into the wind in order to make good the desired track.

wind direction
The direction from which wind is blowing.

wind shear
A change in wind speed and/or wind direction in a short distance, resulting in a tearing or shearing effect. It can exist in a horizontal or vertical direction and occasionally in both.

wind speed
Rate of wind movement in distance per unit time.

wind vane
An instrument to indicate wind direction.

wind velocity
A vector term to include both wind direction and wind speed.

wing loading
The weight of an aircraft per unit wing area.

wing tip vortices   See VORTICES.

words twice
(1) As a request: 'Communication is difficult. Please say every phrase twice'.
(2) As information: 'Since communications are difficult, every phrase in this message will be spoken twice'.

World Area Forecast Centre
A meteorological centre designated to prepare and supply upper-air forecasts in digital form on a global basis to regional area forecast centres.

world area forecast system
A world-wide system by which world and regional area forecast centres provide aeronautical meteorological en-route forecasts in uniform standardised formats.

X

X-axis
(1) The horizontal axis of a two-dimensional Cartesian co-ordinate system.
(2) One of three axes in a three-dimensional Cartesian co-ordinate system.

X-band
A general term of frequencies in the order of 10 000 MHz.

X-channel
A DME or TACAN channel associated or paired with another navaid unit on the same frequency.

X-engine
An engine with four banks of cylinders in the form of an X when viewed from the front or rear.

xenon (Xe)
An inert gas used in some gas discharge tubes as a pumping source for lasers.

X-radiation
Treatment with or exposure to X-rays.

X-ray
Extremely short wavelength EM radiation with a higher frequency than any other but nuclear or gamma radiations.

xylene
Any of three flammable isomeric hydrocarbons of the benzene series. Obtained from wood or coal tar. A mixture of these isomers can be used as a solvent in producing lacquers and as an aviation fuel.

Y

yaw
The rotation of an aircraft about its vertical axis.

yaw angle
The angle between the longitudinal axis of the aircraft and its initial undisturbed direction of flight.

yaw axis
The vertical axis of an aircraft.

yaw damper
A subsystem of an aircraft which automatically senses the onset of yaw and immediately applies a corrective action.

yellow arc
A yellow-coloured arc on the face of an airspeed indicator, the bottom end of which is normally the highest speed at which the aircraft should be flown in turbulent air conditions.

yield point
The point beyond the elastic limit of a material, at which a sudden increase in strain occurs with only a small increase in stress.

yoke
The type of pilot's control column in which aileron control is achieved through the rotational movement of a control wheel/spectacles attached to the top end of the column.

Young's modulus
The ratio of the stress per unit area of cross section on a wire or rod under tension or compression to the longitudinal strain. Also called Young's modulus of elasticity.

Z

zap flap
An early type of split flap fitted to aircraft in which the hinge point moves rearwards as the flap is lowered.

zenith
(1) The point on a celestial sphere that is directly above the observer.
(2) The highest point above the observer's horizon attained by a celestial body.

zero-fuel weight
The maximum permitted takeoff weight of the aircraft minus the total usable fuel weight.

zero gravity
The state of weightlessness; the condition of not experiencing the effects of gravity.

zero lifed
A condition applicable to the aircraft structure, engine, or to specific equipment when it has been rectified and inspected to the extent that the flight time cycle may be restarted.

zero lift
A condition in which a wing generates neither positive nor negative lift. It provides a useful reference point for use in the mathematical analysis of stability and control.

zero lift angle
An angle determined by the amount of camber possessed by the wing.

zero lift line
A line through the trailing edge of an aerofoil and parallel to the relative airflow when the lift is zero.

Z marker
A position marker, transmitting continuously on 75 MHz and generally located at the centre of an LF range or other navigational station to assist the pilot in determining the exact time of arrival over the station.

zonal wind
A west wind; the westerly component of a wind. Conventionally used to describe large-scale flow that is neither cyclonic nor anticyclonic.

Abbreviations, Acronyms and Notation

A

A	amber	AAL	above aerodrome level
A	approach lighting	AAM	air/air missile
A	area	AAM	airline administrative
A	cost of unit weight of fuel		messages
A	suffix indicating local time	A and E	airframe and engine
a	acceleration		mechanic (FAA)
a	equatorial radius of the	AAP	Ageing Aircraft
	earth		Programme (FAA)
a	specific range	AARA	air-to-air refuelling area
a	speed of sound at sea	AAS	advanced automation
	level and standard		system (US)
	temperature	AAS	airport advisory service
a	track angle	AASC	Airworthiness
AA	Airship Association (UK)		Authorities Steering
AA	all after		Committee
AA	Belgian CAA	AATD	Aviation Applied
A/A	air-to-air		Technology Directorate
AAAE	American Association of		(US)
	Airport Executives	AB	air base
AAC	airworthiness advisory	AB	all before
	circular (Australia)	AB	automatic weather
AAC	Aviation Administrative		broadcast
	Communications	ABAC	Association of British
AACA	Aircraft Airworthiness		Aviation Consultants
	Certification Authority	ABC	advance booking charter
AACC	Airport Associations	A-BCAS	active beacon collision
	Co-ordinating Council		avoidance system
AACO	Arab Air Carriers	ABE	aerodrome beacon
	Organisation	ABI	advance boundary
AAD	assigned altitude		information
	deviation	ABM	abeam
AAF	anti-icing fluid	ABN	aerodrome beacon
AAF	army air field	ABNML	abnormal
AAFC	Arab Aviation Finance	ABT	about
	Company	ABV	above
AAG	Airports Authority Group	ABV	repeat in abbreviated form
	(Transport Canada)	AC	advisory circular
AAG	Aeronautical Information	AC	airworthiness circular
	Service Automation	AC	aircarrier
	Group (ICAO)	AC	air conditioner
AAI	angle of approach	Ac	altocumulus
	indicator	A/C	aircraft
AAIB	Air Accidents	A/C	approach control
	Investigation Branch	a.c.	alternating current
	(UK Department of	ACAC	Arab Civil Aviation
	Transport)		Council

ACAP	advanced composite airframe programme	**ACSSB**	amplitude-companded single sideband
ACAP	Aeronautical Chart Automation Project (US)	**ACT**	activate message
		ACT	activation message
		ACT	active
ACARS	airborne communication, addressing and reporting system (ARINC)	**ACTG**	acting
		ACTRAM	Advisory Committee on the Safe Transport of Radioactive Material
ACAS	airborne collision avoidance system	**ACTV**	active
		ACYC	anticyclonic
ACC	area control centre	**AD**	aerodrome
ACCA	Air Charters Carriers Association	**AD**	Airworthiness Directive
		ADA	advisory area
ACCA	Air Courier Conference of America	**ADAPT**	ATS data aquisition processing and transfer
ACCID	notification of an aircraft accident (ICAO)	**ADC**	Aerospace Defense Command (US)
ACC-R	area control-radar	**ADCN**	Aeronautical Data Communication Network
ACD	Aeronautical Charting Division of the National Ocean Service (US)		
		ADCUS	advise customs
ACDB	airport characteristics databank (ICAO)	**ADD**	Soviet long-range bombers
ACE	altimeter control equipment	**ADDN**	addition or additional
		ADEG	ATS Data Exchange Requirements Group (ICAO)
ACE	Association des Compagnies Aériennes de la Communauté Européenne		
		ADEKS	Advanced design electronic key system for communication between ATC controllers
ACFF	air cargo fast flow		
ACFT	aircraft		
ACG	ATS System Concept Working Group (EUROCONTROL)		
		ADELT	automatically deployable emergency locator transmitter
ACJ	joint advisory circular		
ACL	altimeter check location	**ADF**	automatic direction finding
ACLD	above clouds		
ACM	air combat manoeuvre (US)	**ADI**	attitude director indicator
		ADIDS	aeronautical digital information display system
ACN/PCN	aircraft classification number/pavement classification number		
		ADIRS	air data inertial reference system
ACP	Africa, Caribbean and Pacific Group of States	**ADIS**	airport data information system
ACP	air lift command post (US)	**ADISP**	Automated Data Interchange Systems Panel (ICAO)
ACPT	accept, or accepted		
ACR	aerodrome control radar		
ACRBT	acrobatic	**ADIZ**	air defense indentification zone (US)
ACRS	across		
ACSL	altocumulus standing lenticular	**ADJ**	adjacent

ADNL	additional
ADP	aéroports de Paris
ADR	advisory route
ADR	advisory rule
ADREP	accident/incident data reporting system (ICAO)
ADS	address
ADS	automatic dependent surveillance
ADSEL	address selective SSR
ADTS	approved departure times
ADV	advise
ADV	advisory area
ADV	Arbeitsgemeinschaft Deutscher Verkehrsflughäfen
ADVN	advance
ADVY	advisory
ADZ	advise
ADZY	advisory
Ae	aerodrome report (METAR)
AEA	all-electric aeroplane
AEA	Association of European Airlines
AECMA	Association Européenne de Constructeurs de Matériels Aérospatial (European Association of Aerospace Equipment Manufacturers)
AEEC	Airline Electronic Engineering Committee
AEEC	Association of European Express Carriers
AEF	Airfields Environment Federation (UK)
AEIS	Aeronautical En-route Information Service
AEP	Airports Economic Panel (ICAO)
AER	approach end runway
AERA	automated En-route Air Traffic Control (US)
AERADIO	air radio
AERO	Air Education and Recreation Organisation (UK)
AEROSAT	Aeronautical Satellite Council
AES	aerodrome emergency service
AES	aeronautical earth station
AES	Atmospheric and Environmental Services (Canada)
AEW&C	airborne early warning and control
AF	audio frequency
AF-AUX	airport auxiliary field
AFB	Air Force base
AFC	area forecast centre
AFC	automatic frequency control
AFCAC	African Civil Aviation Commission
AFCAS	automatic flight control and augmentation system
AFCS	automatic flight control system
AFCT	affect
A/FD	Airport/Facility Directory (US)
AFI	African Region (ICAO)
AFI	assistant flying instructor
AFI COM/MET RPG	African Region (ICAO) Communications and Meteorological Regional Planning Group (ICAO)
AFIL	air filed flight plan
AFIS	automatic flight inspection
AFIS(O)	aerodrome flight information service (officer)
AFL	above field level
AFMLC	Aeronautical Frequency Management Committee (UK)
AFMS	advanced flight management system
AFN	American forces network
AFO	airport fire officer
AFR	Air Force route
AFRAA	African Airlines Association
AFRATC	African Air Tariff

	Conference	AILS	automatic instrument landing system
AFRC	armed forces reserve center (US)	AIM	air intercept missile
AFRS	American forces radio stations	AIM	Airmen's Information Manual (US)
AFS	aerodrome fire service	AIOA	Aviation Insurance Officers Association
AFS	aeronautical fixed service	AIP	aeronautical information publication
AFS	Air Force station		
AFT	after	AIP	Airport Improvement Program (US)
AFTN	aeronautical fixed telecommunications network	AIR	airworthiness of aircraft
		AIRAC	aeronautical information regulation and control
AFTN	afternoon		
AG	arrester gear	AIRAD	airmen advisory (local only) (US)
A/G	air–ground communication	AIREP	Air report form for reporting position and meteorological conditions in flight
A/G	air to ground radio		
AGA	aerodromes, air routes and ground aids		
AGCS	air–ground communications system	AIS	Aeronautical Information Service
AGL	above ground level	AIS	audio integrating system
AGM	air/ground missile (US)	AISAG	Aeronautical Information Service Automation Group (ICAO)
AGN	again		
AGNIS	azimuth guidance nose-in-stand	AISAP	Aeronautical Information Service Automation Specialist Panel (ICAO)
AGV	avion à grande vitesse (hypersonic transport aircraft)		
		AISG	Aeronautical Information Services Group (EUROCONTROL)
AH	artificial horizon		
AHD	ahead	AIT	advanced instruction technique
AHP	army heliport (US)		
AHRS	attitude and heading reference system	AITAL	Asociación International de Transportes Aévos Latinamericanos
AHS	American Helicopter Society		
AI	attitude indicator	AL	instrument approach and landing charts
AIA	Aerospace Industries Association	ALA	alighting area
AIAA	area of intense air activity	ALAE	Association of Licensed Aircraft Engineers
AIC	aeronautical information circular	ALCE	air lift control element (US)
AICMA	International Association of Aircraft Manufacturers	ALERFA	alert phase
		ALF	aloft
AID	aircraft installation delay	ALF	auxilary landing field
AIDS	aircraft integrated data system	ALG	along
		ALQDS	all quadrants
AIF	attitude instrument flying	ALR	alerting message
AIL	piloted aircraft, countermeasures (US)	ALR	piloted aircraft, countermeasure,

	passive detecting (US)		Commission (ICAO)
ALS	approach light system	ANCAT	abatement of nuisances
ALS	automatic landing system		caused by air transport
ALSF	approach light system		(ECAC)
	with sequenced lights	ANG	Air National Guard (US)
ALSTG	altimeter setting	ANGB	Air National Guard Base
ALT	altitude		(US)
ALTAC	Latin American Freight	ANGR	Air Navigation (General)
	Carriers Association		Regulation (UK)
ALTN	alternate (aerodrome)	ANLYS	analysis
ALTRV	altitude reservation	ANM	AFTN notification
AM	amplitude modulation		message
AMA	area minimum altitude	ANO	Air Navigation Order
AMBEX	African Region (ICAO)	ANP	air navigation plan
	meteorological bulletin	ANR	advanced non-rigid
	exchange scheme		airship
AMC	acceptable means of	ANSA	International Advisory
	compliance		Group Air Navigation
AMC	aerodynamic mean chord		Services (FDR)
	(MAC)	AO	aircraft operator
AMD	amend	AOA	Aerodrome Owners
AMDA	Airlines Medical Directors		Association
	Association	AOA	airport operation area
AME	authorised medical	AOA	at or above
	examiner	AOB	at or below
AMIDS	airport management and	AOC	aerodrome obstruction
	information display		chart
	system	AOC	airline operational control
AMJ	joint advisory material	AOC	Air Operators Certificate
AMMS	automated airport		(UK)
	maintenance system	AOCI	Airline Operators Council
	(AAG Canada)		International
AMR	airport movement radar	AOCS	attitude and orbit control
AMS	aeronautical mobile		system
	service	AOD	above ordnance datum
AMS	air mass		(Newlyn)
AMS	automated manifest	AOE	airport of entry
	system	AOG	aircraft on ground
AMSC	automatic message	AOP	aerodrome operations
	switching centre	AOPA	Aircraft Owners and
AMSL	above sea level		Pilots Association
AMSS	Aeronautical Mobile	AP	airport
	Satellite Service	AP	autopilot
AMSSP	Aeronautical Mobile	Ap	approach lights
	Satellite Service Panel	APA	Accidents in Private
	(ICAO)		Aviation (ECAC) (now
AMST	advanced medium STOL		renamed SAGA)
	transport	APAPI	abbreviated precision
AMSU-B	advanced microwave		approach path indicator
	sounding unit-B	APC	aeronautical public
AMT	amount		correspondence (public
ANC	Air Navigation		telephone)

APC	area positive control	ARINC	Aeronautical Radio, Inc
APCH	approach	ARMET	area forecast, upper
APENN	Aeronautical Fixed		winds and
	Service Planning Study		temperatures at specific
	Group for EUR/NAM/		points
	NAT (ICAO)	ARND	around
APEX	advance purchase	ARO	ATS reporting office
	excursion fare	ARP	airport reference point
APG	piloted aircraft, radar, fire	ARP	air raid precautions
	control (US)	ARP	aviation regulatory
APHAZ	aircraft proximity hazards		proposals (Australia)
API	American Petroleum	ARPT	airport
	Institute	ARR	arrival
APIRG	African Region (ICAO)	ARSR	air route surveillance
	Planning and		radar
	Implementation	ARTCC	air route traffic control
	Regional Planning		centre
	Group (ICAO)	ARTS	automated radar terminal
APP	approach		system
APP	approach control	AS	international scheduled
APPL	aircraft precision position		air transport, alternate
	location equipment		use (aerodromes)
APQ	piloted aircraft, radar,	As	altostratus
	combination of	ASA	air services agreements
	purposes (US)	ASA	Aviation Safety
APP-R	approach control radar		Authorities
APR	April	ASASC	Aviation Safety
APRX	approximate or		Authorities Steering
	approximately		Committee
APS	aircraft prepared for	ASC	assistant sector controller
	service	ASC	automatic systems
APT	airport		controllers
APU	auxiliary power unit	ASCAP	aeronautical satellite
APWI	airborne proximity		communications
	warning indicator		processor
AQZ	area QNH zone	ASCB	avionics standard
A/R	altitude reporting		communications bus
ARB	Airworthiness	ASDA	accelerate–stop distance
	Requirements Board		available
	(UK)	ASDAR	aircraft to satellite data
ARCC	Airworthiness		relay
	Requirements Co-	ASDE	airport surface detection
	ordinating Committee		equipment
ARCP	Aerodrome Reference	ASE	Agence Spatiale
	Code Panel (ICAO)		Europeéne
ARCS	airline request	ASE	altimetry system error
	communication system	ASEAN	Association of South East
ARE	airline revenue		Asian Nations
	enhancement	ASECNA	Agence pour la Sécurité
ARFF	airport/rescue/fire-		de la Navigation
	fighting		Aérienne
ARFOR	area forecast	ASEL	airplane, single engine,

	land (US)		Transporteurs Aériens
ASI	air speed indicator	ATAFG	African Region (ICAO)
ASL	above sea level		Traffic Analysis and
ASM	airspace management		Forecasting Group
ASM	air/surface missile	ATARS	Athinai terminal area
ASM	ad hoc schedule message		radar system (Greece)
ASMA	Aerospace Medical	ATB	automated ticket and
	Association		boarding pass
ASMI	aircraft surface movement	ATC	air traffic control
	indicator	ATC	Air Transport Committee
ASMT	airspace management		(ICAO)
ASP	aircraft servicing platform	ATCA	Air Traffic Control
ASPENN	Aeronautical Fixed		Association (US)
	Service Planning Study	ATCA	air traffic control assistant
	EUR/NAM/NAT	ATCC	air traffic control centre
	Regional Planning	ATCEU	air traffic control
	Group (ICAO)		evaluation unit (UK)
ASPH	asphalt	ATCO	air traffic control officer
ASPP	Aeronautical Fixed	ATC PROSAT	air traffic control project
	Systems Planning for		for satellite
	Data Interchange Panel		communications (ESA)
	(ICAO)	ATCRBS	air traffic control radar
ASR	airport surveillance radar		beacon system
ASR	altimeter setting region	ATCT	air traffic control tower
ASRPG	African Region (ICAO)	ATD	actual time of departure
	SSR Regional Planning	ATD	automatic threat detection
	Group (ICAO)		system
ASSR	airport surface	ATE	air traffic engineer (UK–
	surveillance radar		NATS)
ASTERIX	all purpose structure	ATE	automatic test equipment
	EUROCONTROL radar	ATECMA	Agrupación Técnica
	information exchange		Española de
ASTOVL	advanced short takeoff		Constructores de
	and vertical landing		Material Aerospacial
	aircraft		(Spain)
ASTRA	Applications of Space	ATFM	air traffic flow
	Technology Panel to		management
	Requirements of Civil	ATFM-OPS/AUT	air traffic flow
	Aviation (ICAO)		management
ASW	anti-submarine warfare		procedures and
ATA	actual time of arrival		techniques
ATA	airline tarrif analysis	ATFMU	air traffic flow
ATA	Air Training Association		management unit
	(UK)	ATIS	air traffic information
ATA	Air Transport Association		service
	(US)	ATIS	automatic terminal
ATAA	Air Transport Auxiliary		information service
	Association	ATITA	Air Transport Industry
ATAC	Air Transport Association		Training Association
	of Canada	ATLB	Air Transport Licensing
ATAF	Association		Board (UK)
	Internationale de	ATM	airspace and traffic

	management (ICAO)		(EUROCONTROL)
ATMC	airspace and traffic	**AUTH**	authorised
	management centre	**AUW**	All up weight
ATMG	Airspace and Traffic	**AV**	Abbreviated visual
	Management Group		approach slope
ATN	aeronautical		indicator system
	telecommunications	**AVAD**	automatic voice alerting
	network (SICASP–		device
	ICAO)	**AVASI**	abbreviated visual
ATOA	Air Transport Operators		approach slope
	Association (UK)		indicator
ATOL	air travel organiser's	**AVG**	average
	licence (CAA)	**AVGAS**	aviation gasoline
ATPL	airline transport pilot's	**AVLS**	automatic vehicle location
	licence		systems
ATRP	Air Transport Regulation	**AVMED**	aviation medicine (ICAO)
	Panel (ICAO)	**AVSAT**	Aviation Satellite
ATS	air traffic services		Communications
ATSORA	air traffic services outside		Service
	regulated airspace (UK)	**AVSEC**	Aviation Security Panel
ATSPM	air traffic services		(ICAO)
	planning manual	**AVTUR**	aviation turbine fuel
ATSU	air traffic services unit	**AW**	aerodrome warning
	(UK)	**AW**	airworthiness
ATTN	attention	**AWACS**	airborne warning and
ATU	antenna tuning unit		control system (US)
ATUA	Air Transport Users'	**AWO**	all weather operations
	Association (UK)	**AWOP**	All Weather Operations
ATUC	Air Transport Users'		Panel (ICAO)
	Committee (UK)	**AWOS**	automated weather
AT-VASI	abbreviated TEE visual		observing system
	approach slope	**AWW**	alert weather watch (US)
	indicator	**AWX**	account weather
ATZ	aerodrome traffic zone	**Awy**	airway
AUG	August	**AZ**	azimuth
AUT-ATFM	automation on air traffic	**AZM**	azimuth
	flow management		

B

B	aerodrome or	b	wing span
	identification beacon	BA	braking action
B	beginning of precipitation	BAA	British Airports Authority
B	blue	BAACMIR	British Aerospace Air
B	hourly costs		Combat Manoeuvring
b	polar radius of the earth		Instrumentation Range

BAAEMS	British Association of Airport Equipment Manufacturers and Services	**BERP**	British Experimental Rotor Programme
BACS	beacon collision avoidance system	**BEUC**	Bureau Européen des Unions de Consommateurs
BAeA	British Aerobatic Association	**BFDK**	before dark
		BFN	beam forming network (SATCOM)
BAIR	An advanced information system designed to provide an information service to airport users	**BFO**	beat frequency oscillator
		BFR	before
		BFR	biennial flight review
		BFS	Bundesanstalt für Flugsicherung (FDR)
BALPA	British Airline Pilots Association	**BGA**	British Gliding Association
BAPA	British Aeromedical Practitioners Association	**BGN**	begin, began
		BHAB	British Helicopter Advisory Board
BAPC	British Aircraft Preservation Council	**BHGA**	British Hang Gliding Association
BASI	Bureau of Air Safety Investigation (Australia)	**BHND**	behind
		BHP	brake horsepower
BAST	basic ATC skills trainer (EUROCONTROL)	**BIECA**	Belgium International Express Courier Association
BATA	British Air Transport Association	**BIM**	blade integrity monitor
BAUA	Business Aircraft Users' Association	**BINOVC**	breaks in overcast
		BK	Signal used to interrupt a transmission
BAZ	Austrian CAA		
BBAC	British Balloon and Airship Club	**BKN**	broken
		BL	between layers
BC	back course	**BLC**	boundary layer control
BC	become	**BLD**	build
BCARs	British civil airworthiness requirements	**BLDG**	building
		BLEU	blind landing experimental unit (UK)
bcd	binary coded decimal		
BCFG	fog patches	**BLM**	background luminance monitor
BCH	basic decision height		
BCKG	backing	**BLN**	balloon
BCM	back course marker	**BLO**	below clouds
BCN	beacon	**BLSN**	blowing snow
bco	binary coded octal	**BLW**	below
BCOP	broken clouds or better	**BLZD**	blizzard
BCST	broadcast	**BM**	back marker
BDRY	boundary	**BMAA**	British Microlight Aircraft Association
BEAMA	British Electrotechnical and Allied Manufacturers' Association	**BMDH**	basic minimum descent height
		BNDRY	boundary
BEAMS	British Emergency Air Medical Service	**BNSC**	British National Space Centre
BER	bit error rate	**BNTH**	beneath

BO	boundary lights	BRITE	bright radar indicator tower equipment
BOH	break off height		
BOMB	bombing	BritGFO	British Guild of Flight Operations Officers
BORG	Basic Operational Requirements and Planning Criteria Group (ICAO)	BRK	break
		BRKHIC	breaks in higher overcast
		BRKN	broken
BOVC	base of overcast	BS	broadcast station (commercial)
BPA	British Parachute Association	b/s	bits per second
BPPA	British Precision Pilots Association	BSI	British Standards Institute
		BSU	beam steering unit
BR	mist	BTL	between layers
BRAF	braking action fair	BTN	between
BRAG	braking action good	BTR	better
BRAN	braking action nil	BTU	British thermal unit
BRAP	braking action poor	BV	Bureau Veritas (France)
BRF	short approach desired or required	BWPA	British Women Pilots' Association
Brg	bearing	BYD	beyond
BRG	bearing		

C

C	(ATC) clears	CAB	Civil Aeronautics Board (US)
C	ATC IFR Flight plan clearance delivery frequency	CACC	civil aviation communications centre (UK)
c	compass	CADC	central air data computer
c	coulomb	CADD	computer aided design and drafting
c	speed of electro-magnetic waves	CADIZ	Canadian air defence indentification zone
°C	centrigrade, Celsius		
CA	(ATC) advises	CAEE	Committee on Aircraft Engine Emissions (ICAO)
CAA	Civil Aviation Authority		
CAAC	Civil Aviation Authority of China	CAeM	Commission for Aeronautical Meteorology
CAAd	Norwegian Civil Aviation Authority	CAEP	Committee on Aviation Environmental Protection
CAARC	Commonwealth Advisory Aeronautical Research Council		
		CAF	cleared as filed
CAAS	Civil Aviation Authority of Singapore	CAG	Circulation Aérienne Génerále (France)
CAATS	Canadian automated air traffic system		

CAIP	civil aircraft inspection procedures	**CASB**	Canadian Authority Safety Board
CAIR	Confidential Aviation Incident Reporting Programme (Australia)	**Casevac**	casualty evacuation
		CASOR	Civil Mediator Organisation (UK)
CAMA	Civil Aviation Medical Association (US)	**CAT**	category
		CAT	clear air turbulence
CAMFAX	facsimile transmission network for meteorological charts	**CAT**	Commercial air transport
		CATC	Commonwealth Air Transport Council
CAN	Committee on Aircraft Noise (ICAO)	**CATE**	Co-ordination of Air Transport in Europe
CANAC	Computer Assisted National Air Traffic Control Centre (Belgium)	**CATO**	civil air traffic operations
		CATSE	Capacity of the Air Transport System (ECAC)
CANP	civil air notification procedure(s)	**CAUFN**	caution advised until further notice
CAP	Civil Air Publication	**CAUTRA**	automatic traffic co-ordination
CAP	combat air patrol (US)		
CAP	contact approach	**CAV-OK**	ceiling and visibility OK
CAP	Continuing Airworthiness Panel (ICAO)	**CAVU**	ceiling and visibility unlimited (US)
		Cb	cumulonimbus
CAP	Study Group on Continuing Airworthiness Problems	**CBR**	cloud base recorder
		CBT	computer-based training
		Cc	cirrocumulus
CAPE	convective available potential energy (MET)	**CC**	Co-ordinating Committee (ECAC)
		CC	counter-clockwise
CAPE	computer assisted planning experiment (EUROCONTROL)	**CCC**	Customs Co-operation Council
CAPS	Civil Aviation Purchasing Service (ICAO)	**CCF**	Central Control Function (UK)
CAPSIN	civil aviation packet switching integrated network	**CCIF**	International Telephone Consultative Committee
CAR	Caribbean region (ICAO)	**CCIR**	Comité Consultatif Internationale des Radio communications
CARAT	cargo agents reservation airwaybill insurance and tracking system		
		CCIT	International Telegraph Consultative Committee
CAR	Civil Air Regulations (US)		
CARS	community aerodrome radio station (US)	**CCITT**	Comité Consultatif International pour Télégraphie et Téléphone
CART	cartography (aeronautical)		
CAS	calibrated air speed	**CCTS**	Co-ordination Committee for Telecommunications by Satellite
CAS	close air support		
CAS	collision avoidance system		
CAS	controlled airspace	**CCTV**	closed circuit television

CCW	counter-clockwise	CGAS	coast guard air station (US)
CCZ	coastal confluence zone (US)	CGI	computer-generated image
cd	candela	CGL	circling guidance lights
C_D	drag coefficient	CH	channel
CDB	central data bank (of air traffic demand) (EUROCONTROL)	CH	critical height
		CHAPI	compact helicopter approach path indicator
CDBS	central database system	CHG	change
C_{Di}	coefficient of induced drag	CHIRP	confidential human factors incident reporting system (UK)
CDI	course deviation indicator		
CDP	critical decision point (helicopters)	Ci	cirrus
		CIDIN	common ICAO data interchange network
C_{Dp}	coefficient of profile drag	CIE	Commission Internationale de l'Éclaivage
CEAC	Committee for Central European Airspace Co-ordination		
		CIG	ceiling
CEE	Centre Experimental (EUROCONTROL)	C-in-C	Commander-in-Chief
		CIRM	Comité Internationale de Radio Maritime
CEP	Central East Pacific Ocean Region (ICAO)		
		CIRNAV	circumnavigate
CEPT	Council of European Post and Telegraph	CIS	cargo information system
		CIS	co-operative independent surveillance
CERAP	centre radar approach control		
CES	coast earth station (satellite)	CISPR	Comité International Spécial des Peturbations Radiophoniques
CETD	calculated estimated time of departure		
		CIT	Chartered Institute of Transport
CETO	calculated estimated time of overflight		
		CIT	near or over large towns
CEU	Central Executive Unit (CTMO)	CIV	civil
		CIWS	central instrument warning system
CEV	Centre d'Essais en Vol (France)		
		CL	centreline lights
CFAR	constant false alarm rate	C/L	centre lights
CFCF	central flow control facility	C/L	centreline
		C_L	lift coefficient
CFD	computational fluid dynamics	CLA	clear ice
		CLBR	calibration
CFI	chief flying instructor	CLD	cloud
CFMU	centralised flow management unit	CLL	centreline lighting will be provided
CFN	confirm	C_{Lmax}	maximum lift coefficient
CFO	central forecast office	CLNC	clearance
CFR	Code of Federal Regulations (US)	CLR	clear
		CLRS	clear and smooth
CFR	crash fire rescue	C_{Ls}	lift coefficient at which stall commences
CFS	Central Flying School (UK)		
CG	centre of gravity		

CLSD	closed
CME	Common minimum evaluation exercise of CDB data (EUROCONTROL)
CMPT	common medium term plan (EUROCONTROL)
CMS	common modular simulator (PHARE)
CNL	cancel
CNS	communications navigation and surveillance system
CNS	continuous
CNTR	centre
CNTRL	central
C/O	commanding officer
COCESNA	Central American Corporation for Air Navigation Services
C of A	certificate of airworthiness
C of E	certificate of experience
Coin	counter insurgent
COM	communications
COMCO	computerised operation and maintenance concept
COMIL	Co-ordination Civiles-Militaires (France)
COM/NAV	communications and navigation aids
COMO	compass locator
COM/OPS	Communications/ Operations Divisional Meetings (ICAO)
COMPAS	computer oriented metering, planning and advisory system
COMSAT	Communications Satellite Organisation
COMSND	commissioned
CON	consol beacon
CONC	concrete
COND	condition
CONT	continue
CONUS	Continental United States
COP	change over point
COP	common operational concept (EUROCONTROL)
CORFT	working group on operational requirements impacting on aircraft design (ECAC)
CORR	corridor
COSAC	computing systems for air cargo
COSLANE	constant optimum separation lane
COSPAR	Committee on Space Research
COSPAS	Cosmos Rescue System (USSR satellites)
CP	centre of pressure
CP	command post
CP	constant power
cP	continental polar air (Met)
CPA	closest point of approach
CPF	complete power failure
CPL	commercial pilot licence
CPL	current flight plan
Cpt	clearance
CPU	central processing unit
CQ	general call to all stations
CR	(ATC) requests
CRCO	Central Route Charges Office (EUROCONTROL)
CRDF	cathode-ray direction-finder
CRM	collision risk model
CRM	crew resource management
CRNA	Centre Régional de la Navigation Aérienne (France)
CRS	computer reservation systems
CRS	course
CRT	cathode ray tube
Cs	cirrostratus
CS	communication station
C/S	call-sign
CSB	carrier and sidebands
CSC	chief sector controller
CSDRBL	considerable
CST	coast
CS/T	combined station/tower
CSTMS	customs
cT	continental tropical air (Met)

CTA	calculated time of arrival (US)
CTA	control area
CTAF	common traffic advisory frequency (US)
CTC	contact
CTL	control
CTLZ	control zone
CTMO	Centralised ATFM Organisation (ICAO)
CTN	caution
Ctr	centre
CTR	civil tilt rotor
CTR	control zone
Cu	cumulus
CUF	cumuliform
CUI	Committee on Unlawful Interference
CUS	customs available
CUTE	common use terminal equipment system
CVFR	controlled VFR
CVR	cockpit voice recorder
CVR	cover
CW	clockwise
CW	continuous wave
CWP	Central West Pacific Ocean Region (ICAO)
CWY	clearway
CYC	cyclonic
CZ	control zone

D

D	danger area
D	day
D	drag
D	runway weight bearing capacity, dual wheel
d	differential
DA	danger area
DA	decision altitude
DAAIS	Danger Area Activity Information Service
DABFN	digital adaptive beam forming network (SATCOM)
DABRK	daybreak
DABS	discrete address beacon system
DAC	Danish CAA
DACS	Danger Area Crossing Service
DADC	digital air data computer
DALGT	daylight
DALR	dry adiabatic lapse rate
DAMA	demand assigned multiple access
daN	DecaNewtons
D and D	distress and diversion
DASS	decrease
DATAS	data link and transponder analysis system
DATCO	duty air traffic control officer
DATTS	data acquisition, telecommand and tracking station (SATNAV)
DAUG	Danger Area Users Group (NATS UK)
DAVSS	Doppler acoustic vortex sensing equipment
db	decibels
DBC	Data Bank Comecon
DB COMECON	Data Bank Council for Mutual Economic Assistance
DBE	Data Bank EUROCONTROL
dBm	The relative unit of power compared with a standard of 1mW across an impedence of 600 ohms
DC	direct operating costs
d.c.	direct current
DCA	Directorate of Civil Aviation
DCD	double channel duplex

DCKG	docking		Aviación Civil (Spain)
DCR	decrease	DGP	Dangerous Goods Panel
DCS	double channel simplex		(ICAO)
DCT	direct	DH	decision height
DCTS	data communication	DIF	diffuse
	terminal system	Dir	director
DDM	difference in depth	Dist	distance
	modulation	Dist	district
DDS	data distribution system	DLA	delay
DDT	double dual tandem	DLI	delay indefinite
DEA	Drug Enforcement	DLPU	data link processor unit
	Administration (US)	DLS	data link splitter
DEC	December	DME	distance measuring
DECMSND	decommissioned		equipment
DECR	decrease	DME/P	precision DME
DEG	degree	DMLS	Doppler microwave
DEMIZ	Distance early warning		landing system
	military identification	DMSH	diminish
	zone (US)	DNG	danger, or dangerous
DENEB	fog dispersal operations	DNS	dense
DEP	departure	DNSLP	downslope
DEPCOS	departure co-ordination	DO	ditto
	system (Germany)	DOA	dominant obstacle
DER	departure end of runway		allowance
DERD	display of extracted radar	DOC	designated operational
	data		coverage
DES	data exchange system	DOC	direct operating costs
DES	descend to, or descending	Doc	document
	to	DOD	Department of Defense
DEST	destination		(US)
DETF	data exchange test facility	DOM	domestic
DETRESFA	distress phase	DOP	dilution of position
DEWIZ	distance early warning	DORA	Directorate of Operational
	indentification zone		Research and Analysis
	(US)		(UK)
DF	direction finder	DOT	Department of
D/F	direction finding		Transportation (US)
DFC	Duty Free Confederation	DOT$_p$	Department of Transport
DFDR	digital flight data recorder		(UK)
DFS	detailed functional	DOTS	dynamic ocean track
	specification		system (US)
DFTI	distance from touchdown	DP	deep
	indicator	DP	depart
DFTS	downdrafts	DP	dew point
DFUS	diffuse	DPA	Data Protection Act
DFVR	defense visual flight rule		(CAA)
	(US)	DPC	departure control
DG	director general	DPNG	deepening
DG	directional gyro	DPT	depth
DGAC	Direction Génerale	DR	dead reckoning
	d'Aviation Civile	DR	distaster recovery
	(France)	DRFT	drift
DGAC	Direction General	DRG	during

DRIVE	document review into video entry (UK)
DRSN	drifting snow
DRZL	drizzle
DSB	double sideband
DSIPT	dissipate
DSNT	distant
DST	daylight saving time
DT	dual tandem
DTAM	descend to and maintain
DTG	date-time group
DTp	Department of Transport (UK)
DTRT	deteriorate
DTW	dual tandem wheel landing gear
DUAT	direct user access terminal
DUC	dense upper cloud
DUPE	duplicate message
DUR	duration
DURG	during
DVFR	defense visual flight rules (US)
DVI	direct voice input
DVLP	develop
DVOR	Doppler VOR
DVST	direct viewing storage tube
DW	dual wheel landing gear
DWN	downdrafts
DWPNT	dew point
DZ	drizzle
DZ	dropping zone

E

E	east, or eastern
E	emergency (radio frequency or lighting system)
E	estimated
EA	each
EAA	Experimental Aircraft Association
EAC	expect approach clearance
EADI	electronic attitude director indicator
EAMAC	extension de l'École Africaine de la météorologie et de l'Aviation Civile
EANPG	European Air Navigation Planning Group (ICAO)
EARC	Elimination of Ambiguity in Radiotelephony Call Signs Study Group
EARC	Extraordinary Administrative Radio Conference
EARTS	En-route automated radar tracking system (US)
EAS	equivalent airspeed
EAT	expected approach time
EB	eastbound
EB	Executive Board (JAA)
EBAA	European Business Aviation Association
EBS	electronic beam squint-tracking system (SATCOM)
EC	European Commission
ECA	Economic Commission for Africa (UN)
ECA	Emergency Controlling Authority
ECAC	European Civil Aviation Conference
ECAC/US-CRS	ECAC/United States Working Group on Computer Reservation Systems
ECAM	electronic centralised aircraft monitor
E-CARS	enhanced airline communications and

	reporting system
ECM	electronic countermeasures
ECMT	European Conference of Ministers of Transport (ECAC)
ECS	environmental control system
ECU	European currency unit
ED	emergency distance
EDCT	expect departure clearance time (US)
EDD	electronic data display
EDDUS	electronic data display and update
EDI	electronic data interchange
EDP	electronic data processing
EEC	Eurocontrol Experimental Centre
EEC	European Economic Community
EEE	error
EET	estimated elasped time
EFAS	en-route flight advisory service
EFC	expect further clearance
EFEO	European Flight Engineers Organisation
EFF	effective
EFIR	European flight information region
EFIS	electronic flight instrument systems
EFMS	experimental flight management system (PHARE)
EGASF	European General Aviation Safety Foundation
EGATS	EUROCONTROL Guild of Air Traffic Controllers
EGT	exhaust gas temperature
EHA	European Helicopter Association
EHF	extremely high frequencies
EHSI	electronic horizontal situation indicator
EICAS	engine indication and crew alerting system

EIRP	equivalent isotropically radiated power
EL	elevation
ELBA	emergency location beacon (aircraft)
ELEV	elevation
ELint	electromagnetic
ELM	extended length message
ELR	existing, or environment, lapse rate (temperature)
ELR	extra long range
ELSW	elsewhere
ELT	emergency locator transmitter
EM	class of emission (avionics)
EM	emission
EMBD	embedded (cumulonimbus inside stratus clouds)
emf	electromotive force
EMI	electromagnetic interference
EMRP	effective monopole radiated power (radio)
EMS	emergency medical service
EMSS	experimental mobile satellite system
ENDG	ending
ENE	east north east
ENG	engine
ENRT	en-route
ENTR	entire
ENV	ATS environment of DBE
EOBT	estimated off-block time
EPCO	EANPG Co-ordination Meeting (ICAO)
EPIRB	emergency portable indicating radio beacon
EPNL	effective perceived noise level
EPR	engine pressure ratio
EPROM	erasable progammable ROM
EQPT	equipment
ERA	European Regional Airlines Association
ERCC	Engine Requirement Co-ordinating Committee
ERCC	en-route control centre

EROC	Study Group on En-route Obstacle Clearance Criteria	**EURACA**	European Air Carrier Assembly
EROPS	extended range operation aircraft	**EUR ANP**	European Air Navigation Plan (ICAO)
ERP	effective radiated power	**EUR FCB**	European Frequency Co-ordinating Body (ICAO)
ES	Spanish speaking only		
ESA	electronically scanned array (antenna)	**EUROCAE**	European Organisation for Civil Aviation Electronics
ESA	European Space Agency		
ESCAP	Social Commission for Asia and the Pacific	**EURO-CONTROL**	European Organisation for the Safety of Air Navigation
ESCWA	Economic and Social Commission for Western Asia	**EUROFAR**	European future advanced rotorcraft
ESE	east south east	**EUROFLAG**	European Future Large Aircraft Group
ESIS	European space information system	**EUROPILOTE**	European Organisation of Airline Pilots Associations
ESM	electronic support measures		
ESOC	European Space Operations Centre	**EURPOL**	Intra-European Air Transport Policy (ECAC)
ESRO	European Space Research Organisation	**EUR/RAN**	European Regional Air Navigation Meeting (ICAO)
EST	estimate		
ESTEC	European Space Research and Technology Centre (Netherlands)	**EUR/TFG**	European Traffic Forecasting Group (EUROCONTROL)
ETA	estimated time of arrival		
ETD	estimated time of departure	**ev**	every
		EVD	explosive vapour detector
ETE	estimated time en-route	**EVE**	evening
ETG	electronic target generator (APP — radar)	**EVS**	electro-optical viewing system
ETO	estimated time over significant point	**EW**	electronic warfare
		EX	expect
ETOPS	extended range twin operations	**EXCP**	except
		EXER	exercise, exercising, or to exercise
ETP	estimated time of penetration		
		EXP	expect
ETV	elevating transfer vehicle	**EXTD**	extend, or extending
EUM	European Mediterranean Region (ICAO)	**EXTN**	extension
		EXTRM	extreme
EUR	European Region (ICAO)	**EXTSV**	extensive

F

F	area forecasts (ARFOR)
F	condenser discharge sequential flashing lights/sequenced flashing
F	fixed
F	fuel consumed
f	acceleration
f	frequency
f	fuel consumed in a specified phase of flight
°F	degrees Fahrenheit
FA	Soviet Tactical Air Force
FAA	Federal Aviation Administration (US)
FAC	facilities
FAC	Federal Airports Corporation (Australia)
FAC	final approach course
FAC	Financial Advisory Committee (CAA UK)
FAC	forward air controller
FACSFAC	fleet area control surveillance facility
FAD	fuel advisory departure (US)
FADEC	full authority digital engine (electronic) control system
FAF	final approach fix
FAI	Fédération Aéronautique Internationale
FAL	facilitation of international air services (ICAO)
FAM	family of frequencies
FANS	Future Air Navigation Systems Committee (ICAO)
FAP	final approach point
FAR	Federal Aviation Regulations (US)
FAT	final approach track
FATUREC	Federation of Air Transport User Representatives in the EC
FAX	facsimile transmission

FBL	light (of icing, turbulence, interference or static reports)
FBO	fixed based operator
FBW	fly by wire
FC	funnel cloud
FCB	Frequency Co-ordinating Body (ICAO)
FCC	Federal Communications Commission (US)
FCDM	flow control decision message
FCEM	flow control execution message
FCL	flightcrew licensing
FCRS	flightcrew record system
FCS	flight control system
FCST	forecast
FCU	flight control unit
FDFM	Flight Data and Flow Management Group (ICAO)
FDP	flight data processing
FDP	flying duty period
FDPS	flight data processing system
FDR	flight data recorder, or flight data records
FEATS	Future EUR ATM System Concept Working Group (ICAO)
FEB	February
FEI	field engineering instructions (NATS UK)
Fet	field effect transistor
FETO	free estimated time of overflight
FFAR	folding-fin (or free-flight) aircraft or rocket
FFD	FMS flight data
FFH	for further headings
FG	fog
FGMDSS	future global maritime distress and safety system
FHANG	Federation of Heathrow Anti-Noise Groups

	(UK)		guidance system
FIBS	flight information billing system	**FMP**	flow management position (CTMO)
FIC	flight information centre	**FMPG**	Flow Management
FIC	flying instructor course		Planning Group
FIDS	flight information and display system		(ICAO)
FIGAS	Falkland Islands Government Air Service	**FMS**	flight management system
		FMU	flow management unit
FILG	filing	**FM/Z**	inner (or fan) marker
FIND	flight and information system (UK)		(avionics)
FIR	flight information region	**FNA**	final approach
FIS	flight information service	**FOB**	fuel on board
FISA	Fédération International des Sociétés Aérophilatéliques (automated flight information service)	**FOCA**	Swiss CAA
		FOD	flight operations department (CAA UK)
		FOD	foreign object damage
		FOQNH	forecast regional QNH
		FORNN	forenoon
FITAP	Fédération Internationale des Transports Aériens Privés	**FP**	fuel (petroleum)
		FPEEPM	floor proximity emergency escape path marking
FIW	flight input workstation	**FPL**	filed flight plan
FJ	fuel-jet	**FPM**	feet per minute
FL	flight level	**FPPS**	flight plan processing system
FLD	field		
FLG	falling	**FPR**	flight planned route
FLG	flashing	**FPS**	foot/pound/second
FLIR	forward looking infra-red	**FPX**	fuel petroleum (octane unspecified)
Flo	floodlights		
FLO	Informal Flow Control Meeting, European Region (ICAO)	**FQT**	frequent
		FR	flight refuelling
		F_R	frictional force
FLOE	FLO east (ATMG ICAO)	**FR**	from
FLOW	FLO west (ATMG ICAO)	**FR**	route forecast (ROFOR)
FLR	flares	**FRC**	full route clearance
FLRY	flurry	**FREQ**	frequency
FLT	flight	**FRET**	freezing rain endurance test
FLTCAL	flight calibration procedures		
		FRG	Federal Republic of Germany
FLTCK	flight check		
FLUC	fluctuating, fluctuation, or fluctuated	**Fri**	Friday
		FRMN	formation
FLW	follow	**FRNG**	firing
FLY	fly, or flying	**FROPA**	frontal passage
FM	fan marker	**FROSFC**	frontal surface
FM	frequency management	**FRP**	Fares and Rates Panel (ICAO)
FM	frequency modulation		
FM	from	**FRP**	Federal Radio Navigation Plan (US)
FMA	flight mode annunciation		
FMGS	flight management and	**FRQ**	frequent

FRST	frost
FRTO	flight radio telephony operator
FRZ	freeze
FRZN	frozen
FSC	Flight Safety Committee (UK)
FSDO	Flight Standards District Office (US)
FSG	Flight Study Group (EUR)
FSL	full stop landing
FSS	flight service station (US)
FST	first
FT	fort
FT	feet, foot
FTD	flight training device
FTE	flight technical error
FTHR	further, farther
FTL	flight time limitations

FTLB	Flight Time Limitations Board
FTO	Flying Training Organisation
FTS	flexible track system
FTS	flying training school
FTT	flight technical tolerance
FU	smoke
FU	upper wind and temperature forecast (WINTEM)
FWC	fault warning computer
FWD	forward
FX	fuel unspecified
FZ	freezing
FZDZ	freezing drizzle
FZFG	freezing fog
FZRA	freezing rain

G

G	green
G	ground control
G	Frequency guard (radio)
G	gust
g	acceleration due to earth's gravity
GA	general aviation
GA	go ahead, resume sending
G/A	Ground-to-air
GAATS	Gander automated air traffic system (ICAO)
GADO	General Aviation District Office (US)
G/A/G	ground-to-air, air-to-ground
GAIT	general aviation infrastructure tariff (Australia)
GAMA	General Aviation Manufacturers' Association (US)
GAMPS	Gander automated message processing system (ICAO)

GAMTA	General Aviation Manufacturers' and Traders' Association (UK)
GAPAN	Guild of Air Pilots and Air Navigators (UK)
GARTEUR	Group for Aeronautical Research and Technology in Europe
GASCo	General Aviation Safety Committee (UK)
GAT	general air traffic
GATCO	Guild of Air Traffic Control Officers (UK)
GATT	general agreement on tariffs and trade (ICAO)
GAvA	Guild of Aviation Artists
GAWG	General Aviation Working Group (NATMAC UK)
GCA	ground-controlled approach
GCC	Gulf Co-operation Council
GDOP	geometric dilution of

	position		or water)
GECOT	Group of Experts on Costs and Tariffs	**GNDCK**	ground check
		GND CON	ground control
GEJ	Group of Experts on Jurisdiction (ICAO)	**GNDFG**	ground fog
		GNMS	ground network management system
GEM	graphic engine monitor		
GEN	general	**GNSS**	global navigation satellite systems
GEO	geographic, or true		
GEPTA	Group of Experts on Air Transport Policies	**GOS**	gate operating systems
		GP	glidepath
GERAC	co-ordination of the implementation and operation of the European aeronautical fixed telecommunications services (ICAO)	**GPS**	global positioning system (NAVSTAR US)
		GPU	ground power unit
		GPWS	ground proximity warning system
		Gr	aircraft minimum net climb gradient
GETIDA/RIAC	Group of Experts on Illicit Transport of Drugs by Air (ECAC)	**GR**	hail
		GRACYAS	CAR/SAM Aeronautical Fixed Service Regional Planning Group (ICAO)
GFDEP	ground fog estimated, deep		
GIFAS	Groupement des Industries Françaises Aeronautiques et Spatiales	**GRAD**	gradient
		GRADU	gradual, or gradually
		GRANAS	global radio navigation system (Germany)
GLD	glider	**GRDL**	gradual
GLONASS	global orbiting navigation satellite system (USSR)	**GRE**	ground run-up enclosure
		GRVL	gravel
GMC	ground movement control	**GS**	glide slope
		G/S	ground speed
GMDSS	global maritime distress and safety systems	**GSE**	ground support equipment
GMP	ground movement planning	**GSO**	geostationary orbit
		GTIS	ground-based traffic information system
GMR	ground movement radar		
GMS	geostationary meteorological satellite	**GTS**	global telecommunications system
GMT	Greenwich Mean Time		
Gn	green	**GWT**	gross weight
GND	Above ground	**GWVSS**	ground wind vortex sensing system
GND	ground		
GND	surface of the earth (land		

H

H	high altitude	HI	high altitude
H	non-directional radio beacon	HI	high intensity
h	height	HI	high intensity directional lights
H+	hour plus . . . minutes past	HIALS	high intensity approach light system
H24	continuous operation	HIR	Study Group on Harmful Interference to Radio (ICAO)
H24	24-hour service		
HAA	height above airport	HIRL	high intensity runway edge lights
HAA	Historic Aircraft Association	HJ +	Sunrise to . . . minutes after sunset
HAI	Helicopter Association International	HKIE	Hong Kong Institution of Engineers
HAIL	Highlands and Islands Airports plc (UK)	HL	height loss
HAPI	helicopter approach plate indicator system	HLDG	holding
		HLF	half
HARPS	Heathrow Airport radar processing system (UK)	HLSTO	hailstones
		HLYR	haze layer aloft
HASG	Helicopter Airworthiness Study Group (UK)	Hmr	Homer
		HN	sunset to sunrise
HAT	height above touchdown	HND	hundred
HAZ	hazard	HNM	helicopter noise model (HAI Acoustics Committee)
HBN	hazard beacon		
HC	critical height		
HCAA	Hellenic Civil Aviation Authority	HO	by operational requirements
HDEP	haze layer estimated, deep	Ho	hours for operational requirements
HDF	high frequency direction finding station	HOL	holiday
		HOSG	Helicopter Operations Study Group (UK)
HD FRZ	hard freeze		
HDG	heading	HOSP	hospital aircraft
HDWND	head wind	HP	horsepower
HEL	helicopter	HPA	hectopascals
HELIOPS	Helicopter Operations Panel (ICAO)	HPA	high power amplifier
		HPBW	half power beam width
HELP	helicopter emergency life saving programme	HPZ	helicopter protected zone
		HQ	headquarters
HEMS	Helicopter Emergency Medical Services (UK)	HR	Hear
		HR	here
HERAS	Hellenic radar system	HR	hour
Hf	Heat of combustion of fuel	hr/s	hour/s
		HRZN	horizon
HF	High frequency	HS	during hours of scheduled operations
HGT	height		
HI	high		

HSCT	high speed civil transport (US)	**HVDF**	high and very high frequency direction-finding stations (at the same location)
HSI	horizontal situation indicator		
HST	high speed taxi-way turn off	**HVY**	heavy
HT	high tension	**HWVR**	however
HUD	head-up display	**HWY**	highway
HUM	health and usage monitoring	**HX**	irregular service
		HX	no specific working hours
HURCN	hurricane	**HZ**	haze
HVAC	heating, ventilation and air conditioning	**Hz**	Hertz

I

I	initial approach	**IAOPA**	International Council of Aircraft Owner and Pilot Associations
I	island		
I	unit length on a mercator chart	**IAP**	initial approach procedure
IAAC	International Agricultural Aviation Centre	**IAPA**	International Airline Passenger Association
IAAI	International Airports Authority of India	**IAR**	Intersection of air routes
IAARC	International Administrative Aeronautical Radio Conference	**IAS**	Indicated airspeed
		IASA	International Air Safety Association
IAASM	International Academy of Aviation and Space Medicine	**IATA**	International Air Transport Association
		IB	inbound
IABA	International Association of Aircraft Brokers and Agents	**IBAA**	International Business Aircraft Association
		IBAC	International Business Aviation Council
IAC	intergrated avionics computer	**IBIS**	ICAO bird strike information system
IACA	International Air Carriers' Association	**IBN**	identification beacon
		IC	integrated circuits
IAEA	International Atomic Energy Agency	**ICA**	International Cartographic Association
IAF	initial approach fix		
IAI	Instituto Affari Internationali (Italy)	**ICAA**	International Civil Airports Association
IAL	instrument approach and landing chart	**ICAO**	International Civil Aviation Organisation
IAO	in and out of clouds	**ICAS**	International Council of

	the Aeronautical Sciences		IFPL	Planning Group ICAO flight plan
ICC	International Chamber of Commerce		IFPR	IFPS region
ICCAIA	International Co-ordinating Council of Aerospace Industries Associations		IFPS	integrated initial flight plan processing system
			IFPZ	IFPS zone
			IFR	instrument flight rules
ICE	icing		IFRB	International Frequency Registration Board
ICGIC	icing in clouds		IFSS	international flight service station
ICGICIP	icing in clouds in precipitation		IGA	international general aviation
ICGIP	icing in precipitation			
ICPO-INTERPOL	International Criminal Police Organisation		IGB	intermediate gearbox
			IGIA	Interagency Group on International Aviation (FAA)
ID	identifier, or identify			
IDB	integrated database			
IDEAS	International Data Exchange for Aviation Safety (ICAO)		IGS	instrument guidance system
			ILO	International Labour Organisation
IDENT	identification			
IDS	inclined drive shaft		ILS	instrument landing system
IEC	International Electro-Technical Commission		IM	inner marker
			IMC	instrument meteorological conditions
IECC	International Express Carriers Conference			
			IMDT	immediate
IERE	Institution of Electronic and Radio Engineers		IMechE	Institute of Mechanical Engineers
IF	intermediate approach fix		IMG	immigration
IF	intermediate frequency		IMO	International Maritime Organisation
IFA	International Federation of Airworthiness			
			IMPR	improve, or improving
IFALPA	International Federation of Airline Pilot Associations		IMPT	important
			IMT	immediate
IFAPA	International Foundation of Airline Passenger Associations		IMTA	intensive military training area
			IMTEG	European ILS/MLS Transition Group (ICAO)
IFATCA	International Federation of Air Traffic Controller Associations			
			In	inch
IFATSEA	International Federation of Air Traffic Safety Electronic Associations		INA	initial approach
			INBD	inbound
			INC	in cloud
IFF	identification friend or foe		INCERFA	uncertainty phase
IFIP	International Federation for Information		INCL	include
			INCR	increase
IFLIPS	integrated flight prediction system		IND	wind or landing direction indicator
IFPA	IFPS area		INDC	indicate
IFPG	International Frequency		INDEF	indefinite

Info	information		Instrumentation Group
INMARSAT	International Maritime		(US)
	Satellite Organisation	**IRS**	inertial reference system
INO	Indian Ocean Region	**IRVR**	instrumented runway
	(ICAO)		visual range
INOP	inoperative	**IS**	islands
INP	if not possible	**ISA**	international standard
INPR	in progress		atmosphere
INS	inches	**ISB**	independent side band
INS	inertial navigation system	**ISDN**	integrated services digital
INST	instrument		network
INSTBY	instability	**ISEGLO**	tri-colour light emitting
InstE	Institution of Electronics		diodes
INSTL	install, installed, or	**ISJTA**	intensive student jet
	installation		training area (US)
INSTR	instrument	**ISM**	industrial, scientific and
INT	intersection		medical radio
INTCP	intercept		frequency apparatus
INTER	intermittent	**ISMLS**	interim standard
INTL	international		microwave landing
INTMT	intermittent		system
INTR	interior	**ISO**	International
INTRG	interrogator		Organisation for
INTRP	interrupt, interrupted, or		Standardisation
	interruption	**ISOL**	isolated
INTS	intense	**ISOLD**	isolated
INTSF	intensify, or intensifying	**ISU**	inertial sensor unit
INTSFY	intensify	**ISWL**	isolated single wheel load
INTST	intensity	**ITA**	Institut du Transport
INVOF	in the vicinity of		Aérien
INVRN	inversion	**ITC**	inclusive tour charter
I/O	input/output	**ITF**	International Transport
IOVC	in the overcast		Workers Federation
IPCS	Institution of Professional	**ITU**	International
	Civil Servants		Telecommunications
IPS	integrated power system		Union
IPV	ice on runway	**ITV**	independently targeted
IPV	improve		vehicle
IR	infra-red	**IUAI**	International Union of
IR	instrument restricted		Aviation Insurers
	controlled airspace	**I/V**	instrument/visual
I/R	instrument rating		controlled airspace
IRE	Instrument rating	**IVFRC**	in VFR conditions
	examiner	**IVR**	instrumental visual range
IRIG	Inter Range		

J

J	Joule			decelerometer (Canada)
JAA	Joint Airworthiness Authority	JBI	James brake index (Canada)
JACOLA	Joint Airports Committee of Local Authorities	JCAB	Japanese Civil Aviation Bureau
JALPAS	passenger auto-processing system (Japan)	JCPWG	Joint Certification Procedures Working Group (Europe)
JAN	January	JHN	Japanese Helicopter Network
JAR	joint airworthiness requirements	JNC	jet navigation charts
JAS	joint airmiss section	JPL	jet propulsion laboratory (NASA)
JASA	Joint Airworthiness Steering Committee (Europe)	JSC	Joint Steering Committee (JAA)
Jato	jet-assisted takeoff	JSC	Committee on Joint Support of Air Navigation Services
JAWG	Joint Airmiss Working Group (UK)		
JB	jet barrier	JTSTR	jet stream
J-bar	jet runway barrier	JUL	July
JBD	James brake	JUN	June

K

K	invitation to transmit	KLV	Copenhagen Airport Authority
K	Kelvin		
K	knots	KLYR	smoke layer aloft
k	some constant in any equation	km	kilometre(s)
		kmh	kilometres per hour
KA	start-of-message signal in Morse telegraphy	KOCTY	smoke layer over city
		KPA	kilopascal
KARLDAP	Karlsruhe automatic data processing system (ATCC)	KRM	precision approach system localiser (Eastern Europe)
KDEF	smoke layer estimated, deep	Ksi	thousands of pounds per square inch
KDS	keyboard display station	KT	knots
kg	kilogram(s)	kts	knots
kgs	kilograms	kW	kilo watts
kHz	kiloHertz	kWh	kilo watt-hour
KIAS	knots indicated airspeed		

L

L	cleared to land
L	left
L	licensed (ATC)
L	lift
L	lighted
L	locator
LACAC	Latin American Civil Aviation Commission
LAM	logical acknowledgement message
LAME	licensed aircraft maintenance engineer
LAMS	Light Aircraft Maintenance Scheme (UK)
LAN	inland
LARS	Lower Airspace Radar Advisory Service (UK)
LAT	latitude
LATCA	Los Angeles Traffic Control Area
LATCC	London Air Traffic Control Centre
Latr	locator (compass)
LBA	Luftfahrt Bundesamt (Germany)
LBCM	locator back course marker
LBM	locator back marker
lb(s)	pound(s) (weight)
LCC	life cycle cost
LCD	liquid crystal display
LCG	load classification group
LCL	lifting condensation level
LCL	local
LCM	late change message (ATC)
LCN	load classification number
LCTD	located
LCZ	Localiser (instrument landing system)
LDA	landing distance available
LDA	localiser type directional aid
LDG	landing
LDI	landing direction indicator
LDIN	lead-in light system
LD-SVR	slant visibility meter
LED	light emitting diode
LEGBAC	Limited Exploratory Group on Broadcasting-to-Aeronautical Compatibility
LEN	length
LF	low frequency
LFA	local flying area
LFC	level of free convection
LFR	low frequency radio range
LFV	Swedish CAA
LGT	light
LH	left hand
LI	low intensity omni-directional lights
LIFR	low IFR
LIH	light intensity high
LIL	light intensity low
LIM	light intensity medium
LIM	locator inner marker
LIP	limited installation program (US)
LIRL	low intensity runway lights
LITAS	low intensity two-colour approach slope indicator
LJAO	London Joint Area Organisation (ATC)
LKLY	likely
LLLTV	low light-level television
LLTV	low-light television
LLWS	low level wind shear
LLZ	localiser (instrument landing system) (ICAO)
Lm	lumen
LM	middle locator; middle marker
LMM	middle compass locator; middle marker
LMS	land mobile services
LMSS	land mobile satellite services
LMT	limit
LMT	local mean time

LNA	low noise amplifier		(aircraft systems)
LNDG	landing	LSB	lower sideband
LNG	long	LSF	load sheet fuel
LO	connect me to a peforator receiver	LSQ	line squall
LO	locator at outer marker site	LSSG	Lateral Studies Sub-Group (NAVSEP)
LoA	letter of agreement	LST	printed listing report
LOC	ILS localiser	LT	local time
LOC	localiser (Jeppesen)	LTD	limited
LOC	locator (ICAO)	LTF	landline telephony
LOM	compass locator at outer marker	LTG	lightning
		LTGCC	lightning, cloud to cloud
LOM	locator outer marker	LTGCCCG	lightning cloud to cloud and cloud to ground
LONG	longitude	LTGCG	lightning cloud to ground
LOP	line of position	LTGCW	lightning cloud to water
LPC	linear predictive coding	LTGIC	lightning in clouds
LPDT	low power distress transmitter	LTL	little
		LTLCG	little change
LR	last message received by me was	LTMA	London Terminal Control Area (ATC)
LRC	light reflective capacitor	LTR	later
LRC	long range cruise	LTS	lights
LRCO	limited remote communications outlet	LTT	landline teletypewriter
		LTV	load threshold value
LRG	long range	LUT	local user terminal
LRMTS	laser ranger and marked target seeker	LV	light and variable
		LVA	large vertical aperture
LS	remain well to left side	LVL	level
LS	last message sent by me was (or last message was)	LWC	liquid water content
		LWR	lower
		LYR	layer
LSALT	lowest safe altitude	LYR	layer cloud
LSAS	longitudinal stability augmentation system	lx	lux

M

M	Mach number	m$_a$	mass flow of air
M	magnetic	MAA	maximum authorised altitude (IFR)
M	measured ceiling		
m	metre(s)	MAC	mean aerodynamic chord
M	missing	MAC	message act concellation
M	Monday	MAC	Military Airlift Command (USAF)
m	mass		
m	minutes	MAD	magnetic-anomaly

	detector		altitude
MADAP	Maastricht Automatic	MCAS	Marine Corps air station
	Data Processing System	MCC	Mission Control Centre
MADGE	microwave automatic		(SARSAT)
	digital guidance	MCOQ	multiple choice objective
	equipment		question
MAF	Mission Aviation	Mc/s	megacycles per second
	Fellowship	MCU	modular concept unit
Mag	magnetic	MCW	modulated continuous
MAINT	maintenance		wave
MALS	medium intensity	MDA	master diversion
	approach light system		aerodrome
MALSF	medium intensity	MDA	minimum descent
	approach light system		altitude
	with sequenced	MDD	meteorological data
	flashing lights		distribution
MALSR	medium intensity	MDF	medium frequency
	approach light system		direction-finding
	with runway alignment		station
	indicator lights	MDFY	modify
MAMIS	mandatory modification	MDH	minimum descent height
	and inspection	MDS	mobile data service
	summary	MDT	moderate
MAP	aeronautical maps and	MEA	minimum en-route
	charts		altitude
MAP	missed approach point	MEDA	military emergency
MAPt	missed approach point		diversion aerodrome
MAR	at sea	MEGG	merging
MAR	March	MEHT	minimum pilot eye-height
MARAS	Middle Airspace Radar		over threshold
	Advisory Service (US)	MEL	minimum equipment list
MARS	aeronautical information	MEL	multi-engine, land
	retrieval system (UK)	MEML	memorial
MAS	manual A1 simplex	MER	true height above mean
MASOR	Military Mediator		sea level
	Organisation (UK)	MES	mobile earth station
MAST	multiple aircraft	MET	meteorological
	simulation terminal	Met	Meteorological Office
MATO	military air traffic		(UK)
	operations	METAG	Meteorological Advisory
MATZ	military aerodrome traffic		Group (ICAO)
	zone	METAR	aviation routine weather
MAX	maximum		report in meteorological
MAY	May		aeronautical code
mb	millibars	MEW	mean equivalent wind
MBOH	minimum break off height	MF	mandatory frequency
MBR	marker beacon receiver	MF	medium frequency
MBZ	mandatory broadcast	m_f	mass flow of fuel
	zone	MFA	minimum flight altitude
MC	multiple copy for delivery	MFC	multi-frequency code
	to (number) addresses	MFC/LB	multi-frequency/local
MCA	minimum crossing		battery

MFD	malfunctioning display	
MFV	forward visibility more than . . . miles	
MGCS	meteosat ground computer system	
MGIR	motor glider instructor rating	
MHA	minimum holding altitude	
MHDF	medium and high frequency direction-finding stations (at the same location)	
MHVDF	Medium high and very high frequency direction-finding stations (at the same location)	
MHz	megaHertz	
MI	medium intensity	
MI	mile(s)	
MIA	minimum IFR altitude	
MIALS	medium intensity approach light system	
MID	maritime identification digits (COSPAS–SARSAT)	
MID	middle	
MID	Middle East Region (ICAO)	
MIDN	midnight	
MIFG	shallow fog	
Mil	military	
MIM	minimum	
Min	minimum	
min(s)	minute(s)	
MIPS	million instructions per second	
MIRL	medium intensity runway edge lights	
MIS	missing	
MIT	Massachusetts Institute of Technology	
MKR	marker radio beacon	
Mlnd	maximum landing weight	
MLS	microwave landing system	
MLW	maximum certificated landing weight	
MLWA	maximum landing weight authorised	
MLZ	microwave landing	

	system receiver
MM	middle marker
MMEL	master minimum equipment list
MMI	man–machine interface
MMIC	monolithic microwave integrated circuits (MLS)
MMMM	connect to . . . stations
M_{MO}	maximum operating Mach number
M_{ME}	never exceed Mach number
MNLD	mainland
MNM	minimum
M_{NO}	normal operating Mach number
MNPS	minimum navigation performance specifications (ICAO)
MNR	minimum noise route
MNT	monitor, monitored, or monitoring
MNTN	maintain
MOA	military operations area
MOC	minimum obstacle clearance (required)
MOCA	minimum obstruction clearance altitude
MOCC	Meteosat operations control centre
MOD	Ministry of Defence (UK)
MOD	moderate (icing, turbulence, interference or static reports)
MODE S	mode select beacon system (US)
MOGAS	motor gasoline
MOGR	moderate or greater
MON	above mountains
Mon	Monday
MOPS	minimum operational performance standards
MORA	minimum off-route altitude
MORS	Mandatory Occurrence Reporting Scheme (CAA UK)
MOT	Ministry of Transport (Canada)
MOTNE	meteorological

	operational telecommunications network in Europe
MOV	move
mP	maritime polar air (Met)
MP	maintenance period
MP	manifold pressure
MPDS	multi-purpose display system
MPH	miles per hour
MPS	metres per second
MPS	minimum performance specification
MR	magneto-resistive
MRA	minimum reception altitude
MRB	Maintenance Review Board (UK)
MRG	medium range
MRGL	marginal
MRNG	morning
MRP	Air Traffic Service Meteorological Reporting Point
MRP	machine-readable passport
MRS	medical record system (UK)
MRS	mobile radio service
MRTM	maritime
MRTS	microwave repeater test set
MRU	mountain resue units
MS	maintenance schedule
MS	minus
m/s	metres per second
MSA	minimum safe altitude
MSA	minimum sector altitude
MSAW	minimum safe altitude warning
MSG	flight plan and associated update message
MSG	message
MSL	mean sea level
MSR	message . . . has been misrouted
MSS	mobile satellite service
MSS	mobile satellite systems
MSSR	monopulse secondary surveillance radar
MSTR	moisture

mt	metric tonne
mT	maritime tropical air (Met)
MT	mobile terminal (SATCOM)
MT	mountain
MTA	military training area
MTBF	mean time between failures
MTBO	mean time between outages
MTCA	minimum terrain clearance altitude
MTD	moving target detector, or moving target detection
MTI	marked temperature inversion
MTI	moving target indicator
MTL	minimum triggering level
MTMA	military terminal control area
MTN	mountain
MTPA	mobile transponder performance analyser
MTS	mobile telephone service
MTTR	mean time to repair
MTU	metric units
MTW	mountain waves
MTWA	maximum total weight authorised
MUN	municipal
MURATREC	multiradar track reconstitution
MUTA	military upper traffic control area
MVDF	medium and very high frequency direction-finding stations (at the same location)
MVFR	marginal VFR
MWAA	Metropolitan Washington Airports Authority
MWARA	major world air route area
MWE	manufacturer's weight empty
MWO	meteorological watch office
MX	mixed type of ice formation (white and clear)
MXD	mixed

N

N	engine rotational speed	NASC	National AIS System Centre
N	night		
N	north, or northern	NASP	National Aerospace Plan
n	load factor	NAT	North Atlantic Region (ICAO)
N_1	low-pressure compressor speed on two-spool jet engine	NATMAC	National Air Traffic Management Advisory Committee (CAA; UK)
N_2	High pressure compressor speed on two spool jet engine	NATS	National Air Traffic Services (UK)
N/A	not applicable	NAT/SPG	North Atlantic Systems Planning Group (ICAO)
NA	not authorised		
NAA	National Aeronautic Association (US)	NATSU	Nominated air traffic service unit
NAA	National Airports Authority (UK)	Nav	navigation
NAAS	naval auxiliary air station (US)	Navaid	navigation aid
		NAVSAT	navigation satellite system (ESA)
NAATP	New African Air Transport Policy	NAVSEP	Specialist Panel on Navigation and Separation of Aircraft
NAM	North American		
NAMAS	National Measurement Accredited Services (CAA)	NAVSTAR	navigation satellite timing and ranging system (US)
NAOS	North Atlantic Ocean Station	NB	northbound
NAPOL	North Atlantic Policy Working Group (ECAC)	NBA	National Board of Aviation Finland (Finnish CAA)
NAREFA	North Atlantic reference fares (ECAC)	NBAA	National Business Aircraft Association (US)
NARG	Navigation Aids Deployment Criteria and Area Navigation Working Group (ICAO)	NBFM	narrow band frequency modulation
		NC	no change
		NCC	network control centre
NARL	naval arctic research laboratory	NCS	network co-ordination station
NAS	national airspace system (FAA)	NCU	navigation computer unit
		NCWX	no change in weather
NAS	naval air station	ND	I am unable to deliver message . . . (filing number) addressed to aircraft . . . (identification): please notify message originator
NASA	National Aeronautics and Space Administration (US)		
NASAO	National Association of State Aviation Officials (US)		
		NDB	non-directional radio

	beacon
NDB/L	non-directional beacon, or locator
NDCA	National Dutch Civil Aviation Authority
NE	north east
NEH	I am connecting you to a station which will accept traffic for the station you request
NERC	new enroute centre (NATS UK)
NEXRAD	next generation weather radar
NFDPS	National Flight Data Processing System (ICAO)
NFT	navigation flight test
NFPA	National Fire Protection Association (US)
NGT	night
NIHL	noise induced hearing loss
NIL	None, or I have nothing to send to you
NINST	non-instrument runway
NLA	Study Group on New Larger Aeroplanes
NLR	Nationaal Lucht-en Ruimtevaart- laboratorium National Aerospace Laboratory (Netherlands)
NM	nautical mile(s)
NMC	Network Management Centre (SATCOM)
NMI	nautical mile(s)
NML	normal
NMRS	numerous
NMS	network management system
NNC	non-noise certificated aircraft
NNE	north north east
NNW	north north west
No	number
NOAA	National Oceanic and Atmospheric Administration (US)
NODE	national ATC operational display equipment

NOF	International NOTAM Office
NOISE	National Organization to Insure a Sound- controlled Environment (US)
NONP	non-precision approach runway
NoPT	no procedure turn required or authorised without ATC clearance
NORAD	North American Air Defense Command
NOSIG	no significant change
NOTAM	Notices to Airmen
NOTAR	no tail rotor anti-torque system
NOV	November
NOZ	no operating zone
NP	North Pacific Ocean Region (ICAO)
NPA	Notice of Proposed Amendment (JAR)
NPIAS	National Plan of Integrated Airport Systems (US)
NPRM	notice of proposed rule making (US)
NPRs	noise preferential routes
NR	number
NRW	narrow
Ns	nimbostratus
NSPOL	non-scheduled operations policy (Europe)
NSS	National Search and Rescue Secretariat (Canada)
NTAOCH	notice to air operator certificate holders (CAA UK)
NTE	not to exceed
NTIA	National Telecommunications and Information Administration
NTP	normal (standard) temperature and pressure
NTSB	National Transportation Safety Board (US)
NTZ	no-transgression zone

NW	north west	NWS	national weather service (US)
NWC	naval weapons center (US)	NXT	next
NWP	numerical weather prediction	NYO	not yet operating

O

O/A	on or about	OCC	line engaged
OAAN	Organismo Autónomo Aeropuertos Nacionales (Spain)	OCFNT	occluded front
		OCH	obstacle clearance height
		OCL	obstacle clearance limit
OAC	oceanic area control	OCL	obstruction clearance limit
OAC	Operations Advisory committee (CAA UK)		
		OCL	occlude
OACC	oceanic area control centre	OCLN	occlusion
		OCN1	occasional
OAS	obstacle assessment surface	OCP	Obstacle Clearance Panel (ICAO)
OASIS	oceanic area system improvement study	OCR	occur
		OCR	optional character reader
OAT	operational air traffic	OCS	obstacle clearance surface
OAT	outside air temperature	OCT	October
OAU	Organisation of African Unity	OCTA	oceanic control area
		OCU	operational conversion unit
OB	outbound		
OBI	omni-bearing indicator	ODALS	omni-directional approach lighting system
OBP	on-board processing (SATCOM)		
OBS	human observer	ODAPS	oceanic display and planning system (US)
OBS	observe		
Obs	obstacle lights	OFIS	Operational Flight Information Service (ICAO)
OBS	omni-bearing selector		
OBSC	obscure		
OBST	obstruction	OFP	occluded frontal passage
obstn	obstruction	OFSHR	off shore
OCA	obstacle clearance altitude	OFZ	obstacle free zone
OCA	oceanic area	OGE	out of ground effect
OCA	oceanic control area	OIS	obstacle identification surface
OCA	obstacle clearance altitude		
OCA/H$_{fm}$	OCA/H for the final approach and straight missed approach	OK	We agree, or it is correct
		OLDI	on-line-data interchange
		OLF	outlying field (US)
OCA/H$_{ps}$	OCA/H for the precision segment	OLR	off load routes
		OM	outer marker
OCC	occulting (light)	OMTNS	over mountains

ONC	operational navigation chart	OPS	operational performance standards
ONS	Omega navigation system	OPSP	Operations Panel (ICAO)
ONSHR	on shore	OR	operational requirements
OP	operation, or operate	O/R	on request
OPA	opaque (white type of ice formation)	ORD	order
		ORP	operational readiness platform
OPAC	Working Group on Operations of Aircraft (ECAC)	OSCAR	orbiting satellite carrying amateur radio
OPAL	order processing automated line	OSI	open standard interconnect
OPAS	operational assignment (ICAO)	OSV	ocean station vessel
		O/T	other times
OPC	operational control	OTLK	outlook
oper	operate	OTP	on top
Op. hrs.	operation hours	OTR	other
OPMET	operational meteorological information	OTS	organised track structure
		OTS	out of service
		OVC	overcast
OPN	open, opened, or opening	OVR	over
opn	operation	Ovrn	over-run
OPR	operator, operate, operative, operating, operational	OVRN	over-run standard approach lighting system
OPR	operational preference	OVRNG	over-running
OPS	operation, or operates	OWE	operating weight empty

P

P	paved surface	PA3	precision approach runway Category III
P	personal		
P	port	PABX	private automatic branch exchange
P	pressure		
P	primary frequency	PAC	Pacific Region (ICAO)
P	prohibited area	PAD	packet assembler disassembler
P$_2$	co-pilot		
Pa	Pascal	PADS	Port and Airport Development Strategy (Hong Kong)
PA	passenger address system		
PA	power amplifier		
PA	precision approach lighting system	PAGESAT	one-way message delivery (paging) service
PA1	precision approach runway Category I		
		PALLAS	phased automation of the Hellenic ATC system
PA2	precision approach runway Category II		
		PAM	pulse amplitude

	modulation
PAMC	provisional acceptable means of compliance (ICAO)
PANS	procedures for air navigation services (ICAO)
PANS/OPS	procedures for air navigation services: aircraft operations (ICAO)
PANS/RAC	procedures for air navigation services: rules of the air and air traffic services (ICAO)
PAPA	parallax aircraft parking aid
PAPI	precision approach path indicator
Par	parallel
PAR	precision approach radar
PAR	preferential arrival route
PARL	parallel
PATU	Pan-African Telecommunications Union
PATWAS	pilot's automatic telephone weather answering service
PBCT	proposed boundary crossing time
PBDI	position bearing and distance indicator
PBL	probable
PCA	positive control area
PCA	pre-conditioned air system
PCI	pavement condition index
PCL	pilot controlled lighting
PCN	pavement classification number
PCN	personal communications network
PCPN	precipitation
PCPS	Prodat communication and processing system
PCS	power conditioning system
PCZ	positive control zone (US)
p.d.	potential difference
PD	period
PDAR	preferential departure

	and arrival route
PD(CLC)	pictorical display (course line computer)
PDM	pilot decision making (ICAO)
PDN	public data network
PDR	pre-determined route
PDR	preferential departure route
PE	ice pellets
PE	position error
PEC	position error correction
PELTP	Personnel Licensing and Training panel (ICAO)
PEP	peak envelope power
PER	performance
PERM	permanent
Permly	permanently
PET	point of equal time
PFA	Popular Flying Association
PFCU	power flying control unit
PFD	planned flight data
PFD	primary flight display
PFR	permitted flying route
PH	public holidays
PHA	preliminary hazard analysis
PHARE	Programme for Harmonised ATM Research in Europe (EUROCONTROL)
PHAROS	plan handling and radar operating system
PIAC	peak instantaneous airborne counts
PIANEG	planning of the implementation of an improved AFTN/AFS network (ICAO)
PIB	pre-flight information bulletin
PIB	Public Investments Board (India)
PIBAL	pilot balloon
PIC	pilot in command
PIC	potential icing category (MET)
PIDP	programmable indicator data processor
PIREPS	pilots in-flight hazardous weather reports

PISTON	piston aircraft		authorised to conduct
PJE	parachute jumping exercise		flight tests and ground examinations and sign
PL	position line		Certificates of Test and
PLA	practise low approach		Certificates of
PLASI	pulse light approach indicator system	PPL(XMG)	Experience (UK) A PPL examiner
PLN	flight plan		authorised to conduct
plnd	planned		flight tests and
PM	phase modulation		examinations and sign
PMA	Parts Manufacturing Authority		Certificates of Test for SLMGs and sign
PMB	PHARE Management Board		Certificates of Experience (UK)
PMS	performance management system	PPM	pulse position modulation
PN	prior notice	PPO	prior permission only
PN	pseudo-random noise	PPR	prior permission required
PNI	pictorial navigation indicator	PPSN	public packet-switching network
PNL	perceived noise level	PR	primary radar
PNR	point of no return	PRDS	processed radar display
PNR	prior notice required		system
PO	dust devil	PRES	pressure
POB	persons on board	PRESFR	pressure falling rapidly
Pod	streamlined container	PRESRR	pressure rising rapidly
POH	pilot's operating handbook	PRF	pulse-recurrence frequency
POSN	position	PRKG	parking
PPBE	passenger protective breathing equipment	PRO	Procedure and Requirements
PPI	plan position indicator		Overview Working
PPL(A)	Private Pilot's Licence (Aeroplanes) (UK)	PROB	Group (ICAO) probability
PPL(AS)	Private Pilot's Licence (Airships) (UK)	PROC PROCON	procedure protocol converter
PPL(B)	Private Pilot's Licence (Balloons) (UK)	PROFI	multi-national project financing
PPL(G)	Private Pilot's Licence (Gyroplanes) (UK)	PROG PROg	prognostic, prognosis progress
PPL(H)	Private Pilot's Licence (Helicopters) (UK)	Proj PROM	projection programmable read only
PPL(GR)	A PPL examiner authorised to conduct		memory
	ground examinations	PROP	propeller aircraft
	and sign Certificates of	PROSAT	promotional satellite
	Experience (UK)		project (ESA)
PPL(R)	A PPL examiner	PROV	provisional
	authorised to sign	PRP	pulse repetition period
	Certificates of	PRST	persist
	Experience (UK)	PS	plus
PPL(X)	A PPL examiner	PS	positive value
		PSB	plough, sweeper and

	blower (snow clearance)	PTN	procedure turn
PSBL	possible	PTN	public telephone network
PSDN	packet switched data	PTO	part time operation
	network	PTS	plane transport system
PSG	passage, passing	PTT	post, telegraph and
PSGR	passengers		telephone
psi	pounds per square inch		administrations
PSN	position	PUN	prepare a new perforated
PSP	pierced steel planking		tape for message . . . .
PSR	packed snow on runway	pU/T	pilot under training
PSR	point of safe return	PVD	plan view display (radar)
PSR	primary surveillance	PVL	prevail
	radar	PVOR	precision VHF omni-
PSTN	public switched		directional radio range
	telephone network		(VOR)
PT	procedure turn	PVT	personal verifier terminal
PT	public transport	PVT	private operator
PTC	Personal Technical	PWG	Planning Working Group
	Certificate (NATS UK)		(EUROCONTROL)
PTCHY	patchy	PWI	proximity warning
PTLY	partly		indicator
PTN	portion	PWR	power

Q

Q	engine torque	QFU	magnetic orientation of
Q	squall		runway
q	range factor in Le Breguet	QGH	controller interpreted
	ranger formula		VDF letdown
Q-Band	nomenclature for the	QNE	1013.25 mb (setting for
	35 000 MHz (8 mm)		flight level reading)
	microwave band	QNE	altimeter setting 29.92″
QBI	compulsory IFR flight		Hg or 1013.2 mb
Q-Code	international code used in	QNH	amospheric pressure at
	W/T and RTF		mean sea level
	communications	QSTNRY	quasistationary
QDM	magnetic heading to D/F	QSY	'Q code', notifying
	station		change of radio
QFA	meteorological forecast		frequency
QFE	atmospheric pressure at	QTE	true bearing from D/F
	aerodrome level		station
QFI	qualified flying instructor	quad	quadrant
	(military)	QUJ	true heading to D/F
QFLOW	quota flow control		station

R

R	constant in gas equation	
R	radar	
R	radial	
R	radius	
R	rate of turn	
R	reaction	
R	received	
	(acknowledgement of	
	receipt)	
R	red	
R	resistance	
R . . .	restricted area (followed	
	by identification)	
R	revenue	
R	Roentgen	
R	runway	
R	runway lighting	
(R)	In relation to radio	
	indicates that the	
	spectrum is reserved	
	for aeronautical	
	communications of en-	
	route flights.	
RA	radio altimeter	
RA	rain	
RAA	Regional Airlines	
	Association	
RAC	Rules of the Air and Air	
	Traffic Services	
RACC	Reykjavik Area Control	
	Centre	
RAD	radar	
RAD	radar approach aid	
RAE	Royal Aircraft	
	Establishment	
RAeC	Royal Aero Club (UK)	
RAeS	Royal Aeronautical	
	Society	
RAF	Royal Air Force	
RAFC	regional area forecast	
	centre	
RAFL	rainfall	
RAG	ragged	
RAG	runway arresting gear	
RAI	runway alignment	
	indicator	
RAI	Registro Aeronautico	

Italiano (Italian CAA)

RAIL	runway alignment
	indicator lights
RAL Beacon	runway alignment beacon
	(at distance from
	threshold indicated)
RAM	random access memory
RAMP	radar modernisation
	project
RAN	Regional Air Navigation
	Meeting (ICAO)
RANK	replacement alpha
	numeric keyboard (US)
RAPCON	radar approach control
	(USAF)
RAPID	rapid, or rapidly
RAPSAT	ranging and processing
	satellite
RAR	radar arrival route
RARC	Regional Administrative
	Radio Conference
RAREP	radar weather report
RAS	Radar Advisory Service
RAS	rectified air speed
RAS	replenishment at sea
RASC	Regional AIS System
	Centre
RASH	rain showers
RASN	rain and snow
RASP	Radar Applications
	Specialist Panel
	(EUROCONTROL)
RASS	radar analysis support
	system (EEC)
RATCF	radar air traffic control
	facility (USN)
RB	read back
RB	rescue boat
RBI	radar blip identification
RBI	relative bearing indicator
RBN	radio beacon
RC	reverse course
r/c	rate of climb
RCA	reach cruising altitude
RCAG	remote centre air/ground
RCAG	remote communications
	air–ground

RCC	rescue control centre		Switzerland)
RCC	rescue co-ordination centre	REGR	recent hail
RCF	radiocommunication failure	REIL	runway end identification lights
RCH	reach	REL	relative (direction)
RCL	runway centreline	REMSA	Joint Requirements for
RCLM	runway centreline markings		Emergency and Safety Airborne Equipment, Training and
RCLS	runway centreline lighting system	REP	Procedures (ECAC) reporting point
RCM	basic operational requirement planning criteria and methods of application (ICAO)	REQ RERA RESA RESH	request recent rain runway end safety area recent showers
RCO	remote communications outlet	RESN RESTR	recent snow restrict
RCR	route contingency reserve	RETS	recent thunderstorm
RCR	runway condition reading	REV	revision message
R-CRS	report on course	RF	radio facility
RCV	receive	RFCP	Route Facility Costs Panel
RCVS	receives		(ICAO)
r/d	rate of descent	RFF	fire and rescue equipment
RD	relative density		category
RD	report departing	RFF	Study Group on Rescue
RDARA	regional and domestic air route area		and Fire Fighting (ICAO)
RDCE	radio distribution and control equipment	RFFS	Rescue and Fire Fighting Services (aerodromes)
RDG	ridge	RFI	radio frequency
RDH	reference datum height (for ILS)	RFLG	interference refuelling
RDL	radial	RF NOTAM	Royal Flight NOTAM
RDO	radio	RFSP	replacement flight strip
RDP	radar data processing		printer (US)
RDPS	radar data processing system	RFU	radio frequency/ intermediate frequency
RDR	radar departure route		unit
RDSS	radar determination satellite system (US)	RG	international general aviation, regular use (aerodromes)
RE	recent (of weather phenomena)	RG	range
Re	Reynolds number	RGCSP	Review of the General
REC	receive, or receiver		Concept of (aircraft)
REDZ	recent drizzle		Separation Panel
REF	reference		(ICAO)
REFRA	recent freezing rain	RGD	ragged
REG	registration	RGL	runway guard lights
REG	regular	RGN	region
REGA	Schweizerische Rettungsflugwacht (Air Rescue Guard	RGT RH RH	right relative humidity right hand

RHI	range-height indicator (radar)	**ROC**	rate of climb
RHS	right hand side	**ROCC**	Regional Operators Control Center (US)
RIF	re-clearance in flight	**ROD**	repair on demand
RIN	Royal Institute of Navigation	**ROFOR**	route forecast (aeronautical meteorological code)
RIS	radar information service		
RITE	right (direction of turn)	**ROM**	read only memory
RITs	remote interactive terminals	**RON**	receiving only
		RON	remain overnight
RIV	rapid intervention vehicle	**RP**	report passing
RL	report leaving	**RPC**	Recreational Pilot Certificate (US)
RL	runway (edge) lights		
RLCE	request level change en-route	**RPD**	rapid
		RPFS	radio position fixing system
RLD	Rijksluchtvaartdienst (Dutch CAA)	**RPG**	Regional Planning Group (ICAO)
RLW	Regie der Luchtwegen (Belgian Civil Aviation Administration)	**RPL**	repetitive flight plan (ICAO)
RMCDE	radar message conversion and distribution equipment	**RPL**	Standing Working Group on Regional Plans (ICAO)
RMI	radio magnetic indicator	**RPLC**	replace, or replaced
RMK	remarks	**RPM**	revolutions per minute
RMM	remote maintenance monitoring	**RPPG**	Radar Planning and Policy Group (NATS UK)
RMN	remain		
RMS	precision approach system (Eastern Europe)	**RPRT**	report
		RPS	radar position symbol
		RPS	regional pressure setting
RMS	Royal Meteorological Society	**RPT**	repeat
		RPV	remotely piloted vehicle
RN	Royal Navy	**RQ**	indication of a request
RNAV	area navigation	**RQMNTS**	requirements
RNAV/RNPC	area navigation/required navigation performance capability	**RQP**	request flight plan
		RQRD	required
		RQS	request supplementary flight plan
RNG	radio range		
RNMP	Replacement of the Nautical Mile Panel (ICAO)	**RR**	low/medium frequency radio range station
		RR	report reaching
RNPC	required navigation performance capability (ICAO)	**RRA**	radar recording and analysis equipment
		RRP	runway reference point
RNS	international non-scheduled air transport, regular use (aerodromes)	**RRZ**	radar regulation zone
		RS	international scheduled air transport, regular use (aerodromes)
RO	report over	**RS**	remain well to right side
ROBEX	regional OPMET bulletin exchange scheme	**RSC**	rescue sub-centre
		RSC	runway surface condition

RSG	rising			available
RSITA	International Aeronautical Telecommunications Services Regulations	**RTRD**	retard
		RTRN	return
		RTS	return to service
		RTT	radioteleprinter
RSLS	receiver sidelobe suppression	**RUF**	rough
		RUT	standard regional route transmitting frequencies
RSP	responder beacon	
RSPT	report starting procedure turn	**RV**	radar vector
		RV	RAF rescue vessels
RSR	en-route surveillance radar	**RVA**	radar vectoring area
		RVO	runway visibility by observer
RSS	root sum square	
RSSP	Radar Systems Specialist Panel (ICAO)	**RVR**	runway visual range
		RVRC	runway visual range centre
RSTD	restricted	
R/T	radiotelephony	**RVRR**	runway visual range rollout
RTA	required time of arrival	
RTE	route	**RVRT**	runway visual range touch down
RTCA	Radio Technical Commission of America	
		RVV	runway visibility values
RTD	delayed	**R/W**	runway
RTF	radiotelephony	**RWY**	runway
RTG	radiotelegraphy	rx	receiver for reception only
RTM	revenue ton-mile	**RX**	report crossing
RTOAA	rejected takeoff area		

S

S	runway weight bearing capacity, single wheel			(ECAC)
		SALS	separate access landing system (US)
S	second	
s	Siemens	**SALS**	short approaches light system
S	south, or southern	
S	starboard	**SALSF**	short approach light system with sequenced flashing lights
S	supplementary (frequency)	
S	wing area	**SAM**	South American region
(S)	summertime	**SAM**	surface/air missile
SA	safety altitude	**SAM SAT**	South America South Atlantic (ICAO)
SA	sandstorm, or duststorm	
SA	simple approach lighting system	**SAN**	sanitary
		SAP	soon as possible
SABH	class of radio beacon	**SAR**	search and rescue
SAGA	statistics of accidents in General Aviation	**SARP**	signal automatic radar data processing system

SARPS	standards and recommended practices (ICAO)	**SEA**	South East Asia Region (ICAO)
		SEC	second
SARSAT	search and rescue satellite aided tracking system	**SEC**	section
		SEC	Security Problems Working Group (ECAC)
SAS	stability augmentation system		
SATCC	Southern African Transport and Communications Commission	**SEC**	special event charter flight
		SEL	single-engine, land
		SELA	Latin American economic system
SATCO	senior air traffic control officer		
		SELCAL	selective call system
SATCOM	satellite communications	**SEP**	September
Satnav	Satellite navigation	**SER**	service, served, or servicing
SATSAR	Study Group on Statellite Aided Search and Rescue (ICAO)	**SETD**	scheduled estimated time of departure
		SETR	specific equipment type rating (NATS UK – air traffic engineers)
SAW	surface accoustic wave (SATCOM)		
SB	southbound	**SEV**	severe (of icing, turbulence, etc.)
SBAC	Society of British Aerospace Companies	**SFACT**	Service de la Formation Aéronautique et du Controle Technique
SBC	Sonic Boom Committee (ICAO)		
		SFC	specific fuel consumption
SBO	sidebands only (ILS)	**SFC**	surface
SBTs	self-briefing terminals	**SFERICS**	atmospherics
Sc	stratocumulus	**SFLOC**	synotpic reporting of the location of sources of atmospherics
SCAN	surface condition analyser		
SCATANA	security control of air traffic and air navigation aids		
		s.g.	specific gravity
		SG	snow grains
ScATCC	Scottish air traffic control centre	**SGC**	swept gain control
		SGCAS	Study Group on Certification of Automatic Systems
SCOB	scattered clouds or better		
SCT	scattered		
SCTD	scattered	**SGIT**	special group inclusive tour
Sctr	sector		
SD	standard deviation	**SGL**	signal
SDAU	Safety Data and Analysis Unit (UK)	**SH**	showers
		SHF	super high frequency
SDBY	stand by	**SHFT**	shift
SDD	synthetic dynamic display	**SHLW**	shallow
SDF	simplified directional facility	**SHWR**	Shower
		SI	système international
SDF	step down fix	**SI**	straight-in approach
SDM	system definition manual (INMARSAT)	**SIAPS**	standard instrument approach procedures
SDR	service difficulty reports (US)	**SICASP**	Secondary Surveillance Radar Improvements
SDU	satellite data unit		
SE	south east		

	and Collision Avoidance Systems Panel (ICAO)	**SMK**	and Control Systems smoke
SID	standard instrument departure	**SML**	small
		SMOH	since major overhaul
SIF	selective identification feature	**SMR**	surface movement radar
		SMTH	smooth
SIG	signature	**SMWHT**	somewhat
SIGMA	Computerised system of airport movements management (France)	**SNFLK**	snowflakes
		SNOWTAM	snow NOTAM
		SNSH	snow showers
SIGMET	significant meterological information	**SNW**	snow
		SNWFL	snowfall
SIGWX	significant weather	**SOAP**	spectrographic oil analysis programme
SIMUL	simultaneous, or simultaneously	**SOC**	start of climb
SIRT	status and implementation of ECAC recommendations on technical questions	**SOCATA**	Société de Construction d'Avions de Tourisme et d'Affaires (France)
		SODAR	sound detection and ranging
SITA	Société Internationale des Télécommunications Aéronautiques	**SOIR**	Study Group on Simultaneous Operations on Parallel or Near-Parallel Instrument Runways
SIWL	single isolated wheel load		
SKC	clear sky	**SP**	South Pacific Ocean Region (ICAO)
SKED	schedule, or scheduled		
SL	sea level	**SPA**	schedules planning and analysis (US)
SLAA	Sierra Leone Airport Authority	**SPA**	Seaplane Pilots Association (US)
SLAP	slot allocation procedures		
SLAR	side-looking airborne radar	**SPACDAR**	Specialist Panel on Automatic Conflict Detection and Resolution
SLD	solid		
SLEAT	Society of Licensed Aircraft Engineers and Technologists	**SPAR**	light precision approach radar (France)
SLGT	slight	**SPECI**	special weather report (in aeronautical meteorological code)
SLMG	self-launching motor glider		
SLO	slow	**SPECIAL**	special meteorological report (in abbreviated plain language)
SLP	slope		
SLR	slush on runway		
SLS	side lobe suppression	**SPI**	special position identification pulse
SLT	sleet		
SLW	slow	**SPIR**	single pilot instrument rating
s.m.	statute miles		
SMC	surface movement control	**SPL**	supplementary flight plan
SMG	System Management Group (AAG)	**SPOC**	single SAR points of contact (COSPAS-SARSAT programme)
SMGCS	Study Group on Surface Movement Guidance	**SPORA**	Operational Research and

	Analysis Group (EUROCONTROL)	SSR	secondary surveillance radar
SPOT	spot wind	SSR Mode A	SSR providing coded aircraft identity
SPRD	spread		
SPU	signal processing unit	SSR Mode C	SSR providing coded aircraft altitude
SQ	squall		
SQ	squawk	SSR Mode S	SSR with selective interrogation capability
SQAL	squall		
SQLN	squall line	SSR/RPG	SSR Regional Planning Group (ICAO)
SR	sunrise		
SRA	special rules airspace/area	SST	supersonic transport
SRA	surveillance radar approach	SSW	south south west
		St	stratus
SRE	surveillance radar element	STA	straight in approach
		STAP	Statistics Panel (ICAO)
SRG	Safety Regulation Group (CAA UK)	STAR	standard terminal arrival route
SRG	short range	STAR	studies, tests and applied research (EUROCONTROL)
SRP	slot reference point (ATC)		
SRR	search and rescue region		
SR-SS	sunrise–sunset	STATFOR	Specialist panel on Air Traffic Statistics and Forecasts (EUROCONTROL)
SRV	Surveillance		
SRZ	special rules zone		
SRZ	surveillance radar zone		
SS	sliding scale	STB	stop bars
SS	sunset	STBL	stable
SSA	safe sector altitude	STBY	stand by
SSAA	Swedish Society of Aeronautics and Astronautics	STC	sensitivity time control
		STC	supplemental type certificate
SSAL	sequenced flashing lights	STCA	short term conflict alert system
SSALF	simplified short approach light system with sequenced flashing lights	STD	Indication of an altimeter set to 29.92" Hg or 1013.2 mb without temperature correction
SSALR	simplified short approach light system with runway alignment indicator lights		
		STD	standard time
		STD	subscriber trunk dialling
		Std	standard
SSALS	simplified short approach light system	STDY	steady
		STF	stratiform
SSB	single sideband	STG	strong
SSC	solid state frequency converters	ST-IN	straight-in
		STM	storm
SSE	south south east	STN	station
SSIM	standard schedules information manual (ITAA)	STNA	Service Technique de la Navigation Aérienne (France)
SSM	standard schedule message	STNR	stationary
		STOL	short takeoff and landing
SSPA	solid state power amplifier	STOL	slow takeoff and landing
		STP	standard temperature and

	pressure
strat	stragetic ATFM
STS	status
STT	studies, tests and trials (EUROCONTROL)
SUBJ	subject to . . . .
SUPPS	supplementary procedures (ICAO)
Sv	unit of radiation
SVC	service message
SVCBL	serviceable
SVFR	special visual flight rules

SVR	severe
SVR	slant visual range
.SVRL	several
SW	single wheel landing gear
SW	south west
S/W	surface wind
swg	standard wire gauge
SWY	stopway
SYNOP	special meteorological report
SYS	system

T

T	temperature (degrees Kelvin)
T	terrain clearance altitude (MOCA)
T	threshold lighting
T	thrust
T	transmits only (radio frequencies)
t	time
t	tonne(s)
t	trend landing forecast
°T	true (degrees)
TA	transition altitude
TAB	Technical Assistance Bureau (ICAO)
TABS	telephone automated briefing service
TAC	terminal area charts
tact	tactical ATFM
TAF	terminal aerodrome forecast
TA/H	turn altitude/height
TAIL	tail wind
T & S	turn and slip indicator
TAR	terminal area surveillance radar
TARAD	tracking asynchronous radar data
TARPOL	Tariff Policy (ECAC)
TAS	true air speed
TAX	taxy or taxying

Tax	taxiway lights
TAXI	Taxy and parking facilities chart
TBO	time before overhaul
TC	taxiway centreline lighting
TC	tropical cyclone
TC	Technical Committee
TCA	terminal control area
TCA	transcontinental control area
TCAD	traffic alert and collision avoidance device
TCAS	traffic alert and collision avoidance system
TCC	Technical Co-ordinating Committee
TCH	threshold crossing height
TCN	Transport Canada data processing network
TCTSB	Transport Canada Technical Services Branch (ICAO)
TCTSC	Transport Canada Technical Systems Centre (ICAO)
TCU	towering cumulus
TDA	temporary danger area
TDA	today
TDM	time division mutliplex
TDMA	time division multiple

	access system	TMC	terminal control
TDO	tornado	TMG	track made good
TDR	traffic data record	TML	terminal
TDWR	terminal doppler weather radar	TMPRY	temporary
TDZ	touchdown zone	TMW	tomorrow
TDZE	touchdown zone elevation	TNA	thermal neutron analysis
TDZL	touchdown zone lights	TND	trace narcotics detector
TE	taxiways edge lighting	TNDCY	tendency
TEAM	training equipment and maintenance (NATS UK)	TNEL	total noise exposure level
		TNGT	tonight
		TO	To . . . (place)
		T/O	takeoff
TECR	technical reason	TOAA	Study Group on Takeoff Obstacle Accountability Areas
TEL	telephone		
TEMP	temperature	TOC	top of climb
TEMP	temporary	TOD	time of departure
TEMPO	temporary, or temporarily	TODA	takeoff distance available
TEND	tend, or tending to	TOP	cloud top
TERPS	terminal instrument approach procedures	TOPM	takeoff performance monitor
TET	turbine entry temperature	TORA	takeoff run available
TFC	traffic	TOS	traffic orientation scheme
TGC	travel group charter	TOSA	takeoff space available
TGL	touch and go landing	TOVC	top of overcast
TGS	taxing guidance system	TOW	takeoff weight
TGV	Train à Grande Vitesse (France)	TP	terminal processor
		TP	turning point
Th	Thursday	TPA	traffic pattern altitude
THDR	thunder	TPC	tactical pilotage chart
THK	thick	TPG	topping
THLD	threshold	TPSRS	terminal primary and secondary radar system
THN	thin		
THR	threshold	TR	total reaction
Thr	threshold lights	TR	track
THRFTR	thereafter	TRA	radar transfer of control
THRU	I am connecting you to another switchboard	TRA	temporary reserved airspace
THRU	through	TRACON	terminal radar approach control
THRUT	throughout		
THSD	thousand	TRANS	transition
TIBA	traffic information broadcast by aircraft	TRANS	transmits, or transmitter
		TRANSALT	transition altitude
TIL	until	TRANSLEV	transition level
TIP	until past . . . (place)	TRC	cathode ray tube
TIZ	traffic information zone	TRC	traffic counts and listings
TKH	thick	TRE	type rating examiner
TKOF	takeoff	TREND	Landing forecast of conditions during the two hours after the observation time
TL	transition level		
TLS	target level of safety		
TLWD	tailwind		
TMA	terminal control area	TRGB	tail rotor gearbox

TrM	track magnetic	TT	total time
TRML	terminal	TTS	time to station
TROF	trough	TTT	template tracing technique
TROP	tropopause		
TRRN	terrain	TTY	teletype printer
TRSA	terminal radar service area	Tu	Tuesday
		TURBC	turbulence
TRSB	time reference scanning beam	TURBT	turbulent
		T-VASI	tee visual approach slope indicator
TrT	track true		
TS	thunderstorm	TVE	total vertical error (aircraft separation)
TSFC	thrust specific fuel consumption		
		TVOR	(VHF) terminal omni-range station
TSGR	thunderstorm with hail		
TSHWR	thundershower	TWD	towards
TSMT	transmit	TWEB	transcribed weather broadcast (US)
TSMTR	transmitter		
TSO	technical service order (US)	Twr	tower
		TWR	tower (ATC)
TSO	technical standard order (US)	TWR	tower control
		TWRG	towering
TSO	technical standing order (UK)	TWY	taxiway
		TWYL	taxiway-link
TSQLS	thundersqualls	tx	transmitter, or transmission
TSSA	thunderstorm with sandstorm		
		TXT	text
TSTM	thunderstorm	TYP	type of aircraft
TT	teletypewriter	TYPH	typhoon

U

U	intensity unknown	UDA	upper advisory area
U	unlicensed (ATC)	UDDF	up and down drafts
U	until	U_{DE}	speed of vertical gust
u	unpaved surface	UDF	ultra high frequency direction finding station
UA	air report (AIREP)		
UA	until advised		
UAA	upper advisory area	UFA	until futher advised
UAB	until advised by	UFN	until further notice
UAC	upper airspace centre	UHB	ultra high bypass (jet engine)
UACC	upper area control centre		
UAD	upper advisory route	UHF	ultra high frequency
UAR	upper air route	UIC	upper information centre
UBF	underground baggage facility	UIR	upper flight information region
u/c	under construction	UIS	upper information service

UK	United Kingdom	**UPDFTS**	updrafts
UKAATS	UK advanced air traffic system	**UPR**	upper
		UPS	uninterruptible power supply
UKAIP	UK Aeronautical Information Publication	**UPSLP**	upslope
UKISC	UK Industrial Space Committee	**UPU**	Universal Postal Union
		URD	user requirements document
ULD	unit load device		
ULR	ultra long range	**U/S**	unserviceable
UN	United Nations	**USAF**	United States Air Force
UNAVBL	unavailable	**USB**	Upper Sideband
UNCTLD	uncontrolled	**USBT**	upper surface blowing technique (aircrat wing)
UNDP	United Nations Development Programme	**USCG**	United States Coastguard
		USN	US Navy
UNICOM	aeronautical advisory service	**UTA**	upper control area
		UTC	universal co-ordinated time
UNKN	unknown		
UNL	unlimited altitude	**UUMP**	Unification of Units of Measurement Panel (ICAO)
unlgtd	unlighted		
UNREL	unreliable		
UNRSTD	unrestricted	**UWNDS**	upper winds
UNSTBL	unstable	**UWY**	upper airway
UNSTDY	unsteady		

V

V	design diving speed		control)
V	speed (true airspeed)	**VAL**	in valleys
V	variable	**VAL**	visual approach and landing chart
V	volt		
V₁	speed when an engine failure is assumed to be recognised (decision speed)	**VAN**	runway control van
		VAP	visual aids panel (ICAO)
		VAR	variation
		VAR	visual-aural range
V₂	Takeoff safety speed; initial climb-out speed	**VAS**	vortex advisory system
		VASI	visual approach slope indicator
V₂ₘᵢₙ	Minimum takeoff safety speed		
		VAT	target threshold speed
VA	design manoeuvring speed	**VAW**	Volcanic Ash Warnings Study Group
VA	visual approach	**Vᵦ**	design speed for maximum gust intensity
VA	visual approach slope indicator system		
VAAC	vectored thrust aircraft (advanced flight	**Vᵦ**	design speed for maximum gust

	intensity		static reports)
V_c	design cruising speed	VIP	very important person
	wind component	VIS	visibility
	perpendicular to track	VLF	very low frequency
VCR	visual control room	VLNT	violent
VCRG	VHF Channel	V_{lo}	maximum landing-gear-
	Requirements Group		operating speed
	(ICAO)	V_{LOF}	lift-off speed
V_d	design diving speed	VLR	long range search aircraft
VDC	visual display controller	VLY	volley
V_{df}	demonstrated flight	V_{mc}	minimum control speed
	diving speed		with the critical engine
VDF	VHF directional finding		inoperative
VDF	VHF direction finding	VMC	visual meteorological
	station		conditions
VDGS	visual docking guidance	V_{mca}	minimum control speed
	system		in the air
VDP	visual descent point	V_{mcg}	minimum control speed
VDU	visual display unit		on the ground
V_e	equivalent air-speed	V_{mcl}	minimum control speed
VE	visual exempted		during landing
VER	vertical		approach with all
VERVIS	vertical visibility		engines operating
V_F	design diving speed	V_{mcl-1}	minimum control speed
V_f	design flap speed		during landing
V_{fe}	maximum flap extended		approach with the
	speed		critical engine
VFOP	Visual Flight Rules		inoperative
	Operations Panel	V_{mcl-2}	minimum control speed
	(ICAO)		during landing
VFR	visual flight rules		approach with two
VFRG	Visual Flight Rules Group		critical engines
	(ICAO)		inoperative
VFSS	variable frequency	V_{me}	maximum endurance
	selection system		speed (sailplanes)
VGS	visual guidance system	VMH	visual manoeuvring
V_h	maximum speed in level		height
	flight with maximum	V_{MO}	maximum permissable
	continous power		operating speed or
VHF	very high frequency		mach number
VHF/PTN	VHF radio connection to	V_{mu}	minimum unstick speed
	public telephone	VMU	voice mangement unit
	network	V_n	wind component
VI	variable intensity lights		perpendicular to
V_i	jet velocity		heading
VI/d	best lift over drag speed	Vne	never-exceed speed
VIDS	visual information display	Vno	maximum structural
	systems		cruising speed
V_{le}	maximum landing gear-	VOR	very high frequency
	extended speed		omni-directional radio
VIO	heavy (used in qualifying		range
	radio interference or	VORTAC	VOR and TACAN

	beacons on the same geographical site, i.e. co-located, are termed collectively a VORTAC beacon	**Vslg**	minimum steady flight speed with flaps up and no power being applied Stalling speed at the 1 g breakpoint
VOT	radiated test signal (VOR)	**VSM**	vertical separation mimima
VOT	receiver testing facility (VOR)	**V$_{S0}$**	stalling speed with flaps down and no power being applied
VOX	voice operated transmit		
VP	validation parameter	**VSP**	vertical speed
VPR	voice position reports	**Vsse**	minimum safe single-engine speed (multi-engine aircraft)
VR	rotation speed		
VR	veer		
VR	visual route	**VSSG**	Vertical Separation Study sub Group (NAVSEP)
VRB	voice rotating beacon		
VRB/L	variable	**VSTOL**	very short takeoff and landing
V$_{REF}$	reference landing approach speed, all engines operating	**V$_{TDM}$**	minimum demonstrated threshold speed
V$_{ref}$	Reference speed for final approach	**V$_{Tmax}$**	maximum threshold speed
V$_{FEF-1}$	reference landing approach speed, one engine inoperative.	**V$_{Tmin}$**	minimum threshold speed
VRG	vertical reference gyro	**VTOL**	vertical takeoff and landing
VRP	visual reporting point	**VV**	vertical visibility
V$_S$	Minimum speed in stall or minimum steady flight speed	**V/V**	vertical velocity or speed
		V$_W$	tailwind component
Vs	Stalling speed or minimum steady flight speed at which the aircraft is controllable	**VWS**	vortex wake system
		Vx	best angle-of-climb speed
		Vxse	best single-engine angle of -climb speed (twin)
VSA	by visual reference to the ground	**Vy**	best rate-of-climb speed
		Vyse	best single engine rate-of-climb speed (twin)
VSBY	visbility		
V$_{S1}$	Stalling speed or		

W

W	weight	**WACS**	wide angle collimated display system (simulators)
W	west, or western		
W	white	**WAFC**	world area forecast centre
(W)	wintertime	**WAFS**	world area forecast system
w	watts		
WA	Airmet	**WAR**	printed warning report
WA	word after. . .	**WARC-MOB**	World Administration
WAC	world aeronautical charts		

	Radio Conference of Mobile Service (International Telecommunications Union)
W–A–T	weight–altitude–temperature
WATRS	West Atlantic route system (US)
Wb	weber
WB	word before. . .
WBAR	wing bar lights
WBS	weight and balance measuring system (aircraft equipment)
WC	weather centre (off aerodrome)
WCA	wind correction angle
WD	words of groups
WDI	wind direction indicator
WDLY	widely
WDSPRD	widespread
WEA	weather
WEAA	Western European Airports Association
WECPNL	weight equivalent continuous perceived noise level
Wed	Wednesday
WEF	with effect from
WFP	warm front passage
Wh	white
WHO	World Health Organisation
WI	within
WID	width
WIE	with immediate effect
WILCO	will comply
WINTEM	upper wind and temperature forecast

WIP	work in progress
WIST	Study Group on Low Level Wind Shear and Turbulence (ICAO)
WK	weak
Wkd	weekdays
WKN	weaken
WL	will
WMO	World Met Organisation
WND	wind
WNW	west north west
W/o	without
W/P	way point (RNAV)
WPT	way-point
WRM	warm
WRMFNT	warm front
WRNG	warning
WS	SIGMET on wind shear
WS	weather service
WS	wind shear
WSAS	weather service airport station
WSHFT	wind shift
WSW	west south west
WT	water tank
wt	weight
W/T	wireless telegraphy
WTO	World Tourist Organisation
WTSPT	waterspout
WV	wave
W/V	wind velocity
WVAS	automated wake vortex avoidance system
WW/RGS	Wind shear warning/ recovery guidance system
WWW	world weather watch
WX	weather

X

X	on request
XBAR	crossbar (of approach lighting system)
XC	cross country

XCSRA	cross channel special rules area (UK)
XM	extra marker
Xmits	transmits

XMTR	transmitter	**XS**	atmospherics
XN	intersection	**XTK**	cross-track
XPDR	transponder	**XX**	heavy (of weather)

Y

Y	yellow	**YD**	yard(s)
YCZ	yellow caution zone	**YDA**	yesterday

Z

Z	tower	Z_n	displacement perpendicular to heading due to wind
Z	suffix indicating GMT		
Z	VHF station market at RNG or NDB	**ZOC**	zone of convergence
ZFW	zero fuel weight	**ZRPM**	See MWARA
ZM	VHF station, location (Z) marker	**ZRRN**	See RDARA

γ	specific heat ratio	λ	longitude
Δ	finite difference of	μ	coefficient of friction
δ	ratio of pressure	ρ	density
	relative pressure	Σ	sum of
	element of	σ	relative density
ε	equivalent downwash angle		standard deviation
η	efficiency	ϕ	latitude
$\eta\rho$	propulsive efficiency		inclination of aircraft flight path to horizontal
$\eta\tau$	thermal efficiency		angle of bank
ηo	overall efficiency		runway slope
θ	angle		function
	ratio of absolute temperatures–relative temperature	χ	probability
		ω	earth's angular velocity